Gas Sensor Technologies for Environmental Sensing

Online at: https://doi.org/10.1088/978-0-7503-3159-3

Gas Sensor Technologies for Environmental Sensing

Stefan Palzer
Technical University of Dortmund, Department of Electrical Engineering and Information Technology, Professorship for Sensors, Dortmund, Germany

IOP Publishing, Bristol, UK

ISBN 978-0-7503-3159-3 (ebook)
ISBN 978-0-7503-3157-9 (print)
ISBN 978-0-7503-3160-9 (myPrint)
ISBN 978-0-7503-3158-6 (mobi)

DOI 10.1088/978-0-7503-3159-3

Version: 20250701

IOP ebooks

British Library Cataloguing-in-Publication Data: A catalogue record for this book is available from the British Library.

Published by IOP Publishing, wholly owned by The Institute of Physics, London

IOP Publishing, No.2 The Distillery, Glassfields, Avon Street, Bristol, BS2 0GR, UK

US Office: IOP Publishing, Inc., 190 North Independence Mall West, Suite 601, Philadelphia, PA 19106, USA

To my wife, mi amor Maria and to my beloved kids Elias, Alex, and Lisa.

Contents

Acknowledgments

I would like to acknowledge the exceptional effort and artistic contributions of Olga Ivanova in creating the graphics illustrating the various concepts and methods.

Author biography

Stefan Palzer

Stefan Palzer studied physics in Freiburg, Madrid, and Barcelona. He received his Diploma degree in physics from the University of Freiburg, Germany in 2006. Afterwards he joined Cambridge University, UK, where he was involved in research on experimental quantum optics. He received his PhD degree from Cambridge University, in 2010. From 2011 to 2012, he was a consultant for the German Federal Government in the field of security research. He then joined Jürgen Wöllenstein's Laboratory for Gas Sensors at the Department of Microsystems of the University of Freiburg in 2012 as a Group Leader. In 2017 he became a Capture Talent Fellow and joined the Universidad Autónoma de Madrid, Spain. Since 2020 he has been head of the Professorship for Sensors at TU Dortmund, Germany. His main research topics include semiconducting metal oxide-based sensors and advanced spectroscopic methods. His aim is to apply low-cost approaches in demanding applications, such as climate research and in situ monitoring of industrial processes by utilizing optical and solid-state technologies that have a strong potential for miniaturization.

IOP Publishing

Gas Sensor Technologies for Environmental Sensing

Stefan Palzer

Chapter 1

Introduction

The monitoring of industrial and natural processes is the prerequisite for understanding the rules that govern their behaviour (Dunn 2006, Wallace and Hobbs 2006). This is true in many fields of science and technology, including atmospheric science (Thompson 1998), chemical process engineering (Stephanopoulos 2003), or agriculture (Mukhopadhyay 2012) and the relevant parameters entail physical, chemical, and biological quantities. As a result, reliable and long-term stable measurement technologies are in high demand and research and development efforts are ongoing in this highly multi- and transdisciplinary field, where natural sciences, technology, and data science intersect and collaborate to create new concepts, techniques, and methods for implementing sensing devices and systems (Jackson 2019).

In consequence, the world of measurement technologies is ever expanding in many directions and at the same time the significance and influence of sensors on our daily lives is increasing at an accelerated rate. This is in part due to the confluence of a number of technological capabilities that are currently enabling new applications utilizing smaller, more robust, and lower-cost devices to contribute to a broader and more reliable database. Therefore, it is highly likely that any book on sensors is incomplete and obsolete in terms of sensing capabilities by the time it is published. Nonetheless, the discussion of fundamental transducing mechanisms will retain their validity.

In this regard, chemical sensors are no exception and consequently this contribution does not pretend to cover the whole field comprehensively (Göpel *et al* 1989). Neither will it be possible to cover and include complete up-to-date information about all the latest developments in a rather short introductory text on environmental gas sensing. As a subset of chemical sensors, gas sensors do employ numerous methods to detect and quantify gas species and they feature a vast number of different technologies that are used to determine gas species and measure the respective quantity. The reason for this rich playing field is rooted in the properties of gases, which go beyond the individual behaviour of molecules and

doi:10.1088/978-0-7503-3159-3ch1

include collective as well as chemical behaviour. Still, the fundamental transducing mechanisms that enable the gathering of information about the gas composition and converting it into an electronic signal do give some pointers with respect to their potential in terms of selectivity, sensitivity, and stability. Of course, the vast amount of gases and their different properties make it both challenging and interesting to find ways for solving a plethora of measurement tasks. Especially since for many scenarios the size, price, robustness, and power consumption are equally important.

Therefore, the aim of this book is to focus on the introduction of the most important methods and how they may be implemented. Additionally, some possibilities to construct sensing devices based on a combination of mechanisms is presented. To this end, a concise introduction to some of the most important gas measurement technologies is given. By no means does this text claim to offer a complete survey of all techniques used by sensor manufacturers and researchers in the field, since the sheer vastness of possible interactions between a fluid and a sensing device would require many more pages.

During reading it will become clear that many transducing mechanisms have been employed for many decades and are still used in modern sensing devices. Often it is difficult to establish who was the first to use or discover a particular technique. In any case, ignorance towards historic developments is unintentional. By no means is this book aimed to provide an accurate historic account of the development of chemical sensing technologies. However, each chapter will try to grasp some of the history since it is important to keep in mind that fundamentally new sensing mechanisms are scarce and a broad theoretical background to most sensor technologies presented herein was developed many decades ago. The most important recent changes are due to advancements in miniaturization and improved electronic components for precise readout. This does not take away from the fact that the basic sensing mechanisms have often been understood for a long time. This means that anyone interested in researching, developing, improving, and designing chemical sensors is encouraged to read up on the theory and past developments before getting to work. To this end, this text aims to provide references at the beginning of each chapter as a starting point for further reading, which is followed by an introduction to the most important aspects.

When one looks into the existing texts on gas sensing, most will start with a brief history on the topic. In fact, it is a natural starting point since the development of gas sensing technologies is closely intertwined with the development of the mining industry and the related safety issues. Therefore, gases that are toxic, flammable, or asphixiant, are found on top of the list requiring monitoring. With progressing industrialization, the need for detecting more and more gas species expanded rapidly and as the understanding of the interconnections between health, environment, safety, and process efficiency deepened, the imperative for providing reliable data on the gas compositions grew stronger. Even today, the motivation for expanding gas sensing capabilities oftentimes stems from the motivation of preventing toxic, explosive, or otherwise health-adverse situations.

This puts the development of gas sensing capabilities in an interesting perspective: unlike the development of the LASER, which was a solution seeking a problem

(Maiman 1964), the progress in gas detection technology has mostly been and continues to be mainly need-driven. Today, this need manifests itself e.g. when trying to better understand climate change, improve air quality, detect diseases early, optimize agricultural and industrial processes, or ensure the safety of the environment and human health.

Those are some of the current topics that are in need of suitable gas sensing solutions and it is likely that future solutions will employ one or several of the methods presented in this text. However, the concrete implementation will very likely see one or another trick applied in order to provide a satisfactory and reliable solution. So, on the one hand, this book gives a first starting point for scientists and engineers concerned with gas sensor development that are new to the field. On the other hand, users of gas sensing technology should also be aware of the limitations of the technology they intend to deploy in their respective application. After all, a company writing a data sheet may have forgotten to mention the occurance of a cross-sensitivity to other gas species or the dependence of the sensitivity on other parameters.

Historically, gas sensor development started from safety-related applications and evolved in parallel to industrialization and the development of the chemical and pharmaceutical industries. The capability to determine the presence of a certain gas species also opened up new possibilities in other fields, that require chemical information. Today, new technological capabilities in terms of miniaturization and the level of control over light and matter also inseminate new applications. Still the role of selectivity remains a central point in most research and development efforts. Some of the possible application scenarios for gas sensors are symbolized in figure 1.1.

Figure 1.1. The range of possible applications for gas sensors is vast and encompasses environmental research, industrial sites, medicine, agriculture, food and beverages, and air quality monitoring. Each application has a specific set of requirements that gas sensors need to comply with.

Before starting with the deed, no book on sensors can go ahead without a couple of definitions and basic vocabulary that will be needed throughout the book and especially when applying the knowledge in practice.

1.1 Definition

Certainly, first in that list is the definition of a sensor: a sensor is a device converting information about the environment into an electrical signal (Fraden 2016). This is true for all sensors and it is important to keep in mind that the final goal of any gas sensor development has to be a device that outputs an electronic signal. This may be achieved directly, i.e. by converting the environmental parameters into an electronic signal, S. It may also be achieved by using a series of transducing mechanisms that finally end up generating an electronic signal. Without this conversion into an electronic signal one still may achieve building a measurement system but not necessarily a sensor. Lastly, the interpretation of a sensor's signal and the calibration of sensors will not be dealt with here.

To exemplify the different concepts, temperature is a suitable parameter: if you aim to determine, say, your body's temperature, you can use a liquid thermometer. This is, however, not a sensor since you merely look at the height of a liquid, whose volume is temperature dependent, which results in different heights in a capillary. The manufacturer of the device painted a scale onto the glass capillary and one determines the temperature by estimating the expansion of the liquid with respect to that scale. A sensor, on the other hand, has the task to convert that change in liquid expansion into an electrical signal, i.e. a voltage or current. This implies the need to utilize at least one more transducing mechanism that is able to convert the liquid height into an electronic signal. If one wants to build a temperature sensor, it might therefore be easier to choose a different transducing mechanism like, say the resisitivity of a metal, which is usually temperature dependent. This way, by applying a constant voltage to the metal, a temperature-dependent current will occur and you would have achieved your goal, i.e. converting an environmental parameter (temperature) into an electrical signal (current). In gas sensing the fundamental task is no different and the tools available to achieve a conversion of chemical information into an electric signal are at the core of this text.

1.2 Sensitivity

Secondly, the sensitivity $S(x)$ near an input signal x is defined as the ratio of the incremental change in sensor response $\partial R(x)$ and that of the incremental change of the input signal ∂x:

$$S(x) = \frac{\partial R(x)}{\partial x}. \tag{1.1}$$

Figure 1.2 depicts an exemplary relation between sensor response $R(x)$ and input signal to illustrate the meaning of sensitivity. By no means is this a linear relationship over a range of input parameters per se. On the contrary, in almost all instances this relation is a non-trivial, non-linear relation when evaluating it along a range of

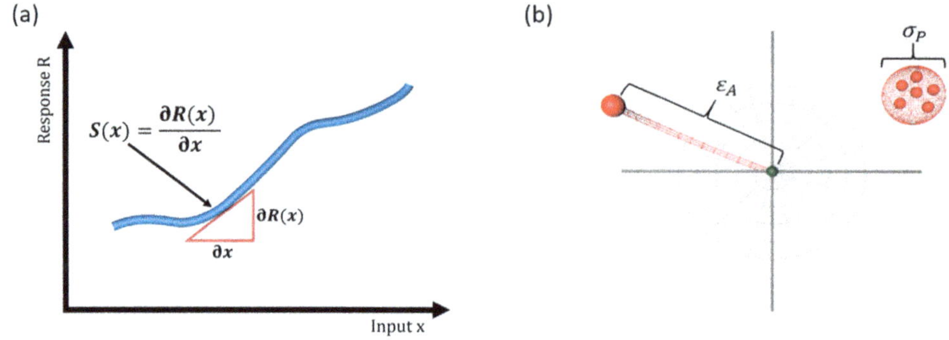

Figure 1.2. (a) Schematic depiction of the definition of the sensitivity: The change in output for a defined change in input is called the sensitivity. This means that for different input values, the sensitivity around this point usually differs. (b) A visualization of accuracy and precision: the true value of the measurand is symbolized by the centre of the spider web. The distance between the centre and the sensor reading is called accuracy, while the scattering of sensor readings for the same input value is called precision.

input values. There is no need to delve too deeply into this topic at this point, since dedicated books on the topic are readily available (Göpel *et al* 1989).

The quality of sensors is determined by its accuracy and precision, which are fundamental for assessing the performance of a sensor. Ultimately, the worth of a sensor reading is determined by the level of both precision and accuracy in unison. Both measures are schematically depicted in figure 1.2.

Accuracy describes how close the measured value is to the true value (BIPM 2012). At this point there will be no philosophical discussion about what truth is. In practice, the national and international institutes for standardization have developed sophisticated reference systems and are able to provide objective, verifiable reference systems. The distance of a sensor reading from the true value is referred to as accuracy and the more accurate a sensor, the smaller that distance is. It is the systematic error, bias or offset of the sensor device and sensor calibration methods aim at optimizing accuracy.

Precision, on the other hand, is a measure of the magnitude of the random error associated with each determination of an input parameter (BIPM 2012). If the sensor is presented with the same situation, the output will vary for each measurement in a statistical, random manner. The smaller this variation is, the higher the precision of the sensor.

It becomes immediately clear that the smallest quantity a sensor can reliably detect is closely related to the presented concepts. In particular, the precision in combination with the sensitivity near zero input will determine the so-called limit of detection, which is one of the key parameters limiting the application scenarios of a gas sensor. Likewise, the input range a sensor can cover is a function of its response function and precision.

1.3 Selectivity

Thirdly, selectivity is a crucial parameter for most scenarios. In various senses it is the most important parameter and also one that is often overlooked by end-users.

Sensor manufacturers will be happy to provide prospective customers with a limit of detection, power consumption, or sensitivity functions for the gas the sensor was made for. However, the cross-sensitivities are often dealt with in less detail due to the possibly large parameter space. They occur as the sensor responds to input parameters x_i other that the intended one x_1, i.e. the response $R(x_1, x_2,\ldots, x_n)$ depends on multiple input parameters x_i (Zipser 1993). Consequently, the sensitivity may no longer be described by equation (1.1) but becomes:

$$S(x_1,\ldots, x_n) = \sum_i S_{x_i} = \sum_i \frac{\Delta R(x_i)}{\Delta x_i}, \tag{1.2}$$

where x_i represents all parameters the sensor responds to. This includes the sensitivity to parameter x_1 the sensor is it intended to determine as well as all cross-sensitivities to x_i with $i \neq 1$, i.e. the cross-sensitivities.

Apart from the dependence of sensors on temperature and other physical parameters, a particular focus for gas sensors has to be placed on the reaction towards other gas species. Ideally, a sensor for a certain gas would only react to one specific input. In practice, however, the number of possible cross-sensitivities is large and one key issue in gas sensing is improving sensor performance in terms of selectivity.

A sensor without any cross-sensitivities is considered specific, i.e. it only responds to the input parameter it has been designed for (Kaiser 1972) $\sum_{i\neq 1}|S_{x_i}| = 0$. To quantify selectivity, different measures have been proposed and it may be expressed as the ratio of sensitivity and the total sum of all cross-sensitivities via the specificity (Danzer 2001):

$$\mathrm{spec}(x_1/x_2,\ldots, x_n) = \frac{S_1 \cdot x_1}{\sum_{i\neq 1} S_i \cdot x_i}. \tag{1.3}$$

Since gas sensors may be exposed to an arbitrary large number of possible gas species a complete quantification is difficult. Therefore, the book's chapters are ordered roughly as a function of the degree of selectivity and a short evaluation of the level of selectivity is added to each technology. Crucially, selectivity is often the determining factor for the usability of a certain gas sensing approach in a defined scenario.

1.4 Stability

Next on the list is the stability of a sensor, which describes the constancy of the sensor reading with time. One obvious factor influencing the stability is ageing of the sensor's components and the associated drift in output signal. Additionally, external factors including vibrations or dirt may affect the sensor performance in various ways. Importantly, a lack of stability may manifest itself in numerous ways, including a change in sensitivity, selectivity, or baseline signal. The strategy to tackle a lack in stability is oftentimes twofold: first, mitigating the effects on sensor performance may be achieved via re-calibration of the sensor or employing models that predict changes of the sensors' output with time and correct for it. The latter may lead to sophisticated approaches that significantly expand the lifetime of sensors. Second, adjusting the construction of the sensor itself, e.g. by including reference channels, or more robust components and designs.

In summary, the most fundamental parameters to characterize a sensor are sensitivity, selectivity, and stability, which are sometimes referred to as the 3S. Improvement of these parameters is a constant goal for engineers and researchers alike and competing approaches usually do not perform equally on all three domains.

This book is structured in such a way as to enable the reader to understand the working principles of most of the important gas sensors classes and judge the merits and shortcomings of different gas sensing technologies by themselves afterwards.

In consequence, the text starts with the most important properties of matter in gaseous form. It is a very brief account and not a lot of pages are spent on deriving those properties from first principles. Instead, readers are directed and referred to textbooks on statistical physics, chemistry, engineering, and spectroscopy if deeper insights are required or the brief explanations given here are unsatisfactory.

In the second part of the book, transducing mechanisms employed in gas sensing are presented. To give the presentation a certain degree of logic and ordering, the techniques are described starting from the least selective methods to the most selective ones. This means that sensing approaches that are able to detect a gas with certain properties but not distinguish between gas species that possess this property are presented first. As the text evolves, the capabilities to narrow down the gas species improve. As a top-up, combined systems are briefly presented, that aim to combine selectivity and sensitivity and allow for identifying and quantifying a gas species. Figure 1.3 provides an overview of the gas sensing methods dealt with in terms of their selectivity assessment, which also provides a means for structuring this text.

The attempted ordering does not consider further performance indicators, which may include, e.g., the capability to selectively detecting various gases. Most certainly there are a number of exceptions from the ordering presented in figure 1.3 but it will form the guiding structure of this introductory text nonetheless.

Apart from the necessary sensor performance data, the measuring conditions set the framework for the possible technological realisation. Especially in industrial settings, high temperatures and pressures often rule out the use of particular technology classes. The types of gases that need determination are equally disparate and may range from small molecules to long hydrocarbons. Ultimately, microscopic and macroscopic properties of each gas species determine possible routes to their detection. While each chapter intends to provide examples of applications, each measurement task will require its own assessment regarding the suitability of a sensing approach.

As stated earlier, the idea behind the structure of this text is for the reader to get to know the most widely used transducing mechanism to detect gases and evaluate the respective potentials in terms of stability, selectivity, sensitivity, and miniaturization. The following chapters are aimed at providing the necessary background to

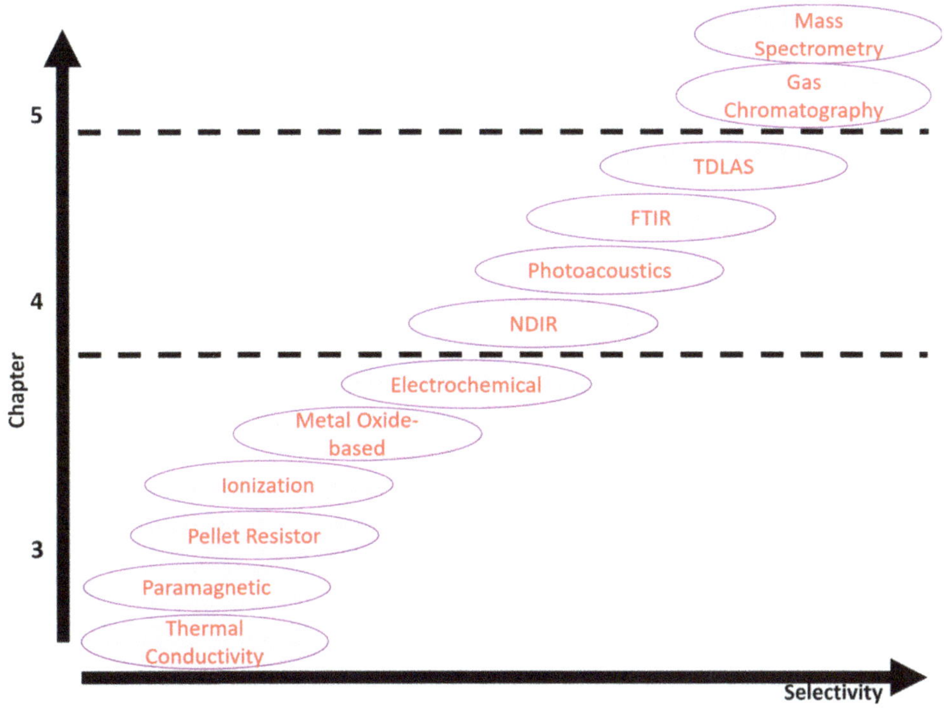

Figure 1.3. An overview of the different gas sensing techniques that are discussed in this book including the respective chapter numbers they are presented in. The assessment is based on the underlying transducing mechanisms and the actually achieved selectivity does depend on the implementation of the sensor and is situation dependent, i.e. the evaluation of selectivity is conditional on the measurement task.

understanding the working principles of the different gas sensing technologies and to give pointers for more in-depth reading.

Chapter 2 provides the background for most detection technologies and will give a brief introduction to those microscopic and macroscopic properties of gases that give rise to possible transducing mechanism for their detection and quantification. It uses concepts from classical and quantum mechanics, without deducing the respective results. Further theoretical background on the respective properties are readily available elsewhere and the chapter is more of a reminder of some basic concepts than a description of the theoretical background.

Chapter 3 is mainly concerned with direct transducers, i.e. sensing approaches that are able to directly convert chemical information into an electronic signal. Those technologies only require electrodes to read out the gas-induced changes in electronic properties. In a sense, chapter 3 is therefore concerned with chemical sensors in its literal, original meaning. Nonetheless, the lines between direct and indirect chemical sensors quickly becomes blurred and as a consequence, no clear line has been drawn within chapter 3. As a consequence of the transducing mechanisms deployed for the gas sensor types presented in this chapter, the selectivity is typically the lowest, as is shown in figure 1.3.

Chapter 4 is concerned with spectroscopic methods, which in this sense are definitely indirect chemical sensors, since they rely on the interaction between light and matter to extract chemical information. The transducing is indirect, since first light is altered by interacting with gas molecules, and afterwards the light itself is analyzed. The spectroscopic methods kick off with a second, short theoretical part to introduce the basic concepts of light–matter interaction. Afterwards, some of the more important techniques to make use of the various forms of interaction are presented. A particular focus lies on absorption spectroscopy tools, since the most widely used and versatile techniques are based on absorption of light by gases. The potential to construct highly selective gas sensors for a broad range of gas species has led to a proliferation of spectroscopic tools and further developments are likely to bring further advances in terms of miniaturization and sensitivity.

In chapter 5 complex gas detection systems will be introduced briefly. The idea behind those systems is to combine a mechanism performing the separation of molecules with a sensitive gas molecule detector. Usually, the latter is non-selective but allows for a quantitative determination of the number of molecules. In this way, a highly selective and sensitive determination of complex gas matrices becomes possible. However, the necessary infrastructure to operate those devices implies an inherent difficulty in miniaturizing those devices. The techniques employed to quantify the gases are presented in chapter 3.

In each chapter a couple of exemplary applications for various types of gas sensors are described. Since the use-cases for gas sensors cover such a large spectrum, the idea behind using sample applications is to illustrate how a requirement analysis has led to the respective design considerations and finally the implementation. After all, the choice for a particular approach is influenced by multiple parameters and the mere functionality is often not the decisive factor in decision making.

References

BIPM 2012 *JCGM 200:2012 International Vocabulary of Metrology—Basic and General Concepts and Associated Terms (VIPM)*

Danzer K 2001 Selectivity and specificity in analytical chemistry. General considerations and attempt of a definition and quantification *Fresenius' J. Anal. Chem.* **369** 397–402

Dunn W C 2006 *Introduction to Instrumentation, Sensors and Process Control* (Artech House Sensors Library) (Norwood, MA: Artech House)

Fraden J 2016 *Handbook of Modern Sensors: Physics, Designs, and Applications* 5th edn (Cham: Springer International Publishing)

1989 *Sensors: A Comprehensive Survey* ed W Göpel, J Hesse, J N Zemel, T Grandke and W H Ko (New York: VCH)

Jackson R G 2019 *Novel Sensors and Sensing* (Boca Raton, FL: CRC Press)

Kaiser H 1972 Zur Definition von Selektivitt, Spezifitt und Empfindlichkeit von Analysenverfahren *Anal. Bioanal. Chem.* **260** 252–60

Maiman T 1964 Developer of the laser calls it 'A solution seeking a problem'; President of Korad spends spare time gardening and fixing TV sets *New York Times* May https://www.nytimes.com/1964/05/06/archives/developer-of-the-laser-calls-it-a-solution-seeking-a-problem.html

Mukhopadhyay S C (ed) 2012 *Smart Sensing Technology for Agriculture and Environmental Monitoring* (Lecture Notes in Electrical Engineering vol 146) *(Berlin: Springer)*

Stephanopoulos G 2003 *Chemical Process Control: An Introduction to Theory and Practice* (Prentice-Hall International Series in the Physical and Chemical Engineering Sciences) (Englewood Cliffs, NJ: Prentice-Hall)

Thompson R D 1998 *Atmospheric Processes and Systems* (Routledge Introductions to Environment Series 1st edn (Taylor and Francis Group)

Wallace J M and Hobbs P V 2006 *Atmospheric Science: An Introductory Survey* (International Geophysics Series vol 92) 2nd edn (Amsterdam: Elsevier)

Zipser L 1993 Selectivity of sensor systems *Sens. Actuators* A **37–8** 286–9

IOP Publishing

Gas Sensor Technologies for Environmental Sensing

Stefan Palzer

Chapter 2

Properties of gases

Sensing devices may utilize different properties of a gas or gas mixture as a transducing mechanism. In order to be able to evaluate the potential of sensing devices for particular applications it is important to understand the fundamental principles that govern a gas's behaviour. Consequently, gas sensor performance is influenced by a multitude of effects that may be interdependent and lead to cross-sensitivities. This chapter will summarize the most important properties that are of relevance for the performance of gas-detecting devices and which ultimately hold the key for evaluating the selectivity, sensitivity, and stability of gas sensors.

A gas is a compressible fluid that will expand to fill its container (Goodstein 2014). This makes it fundamentally different from liquids, which have a fixed volume, and solids, which will retain their volume as well as their form even without a container. Temperature and pressure mainly determine the structural phases of matter and so-called phase diagrams show the state as a function of these two parameters. Even in liquid or solid state most matter exhibits a non-zero vapour pressure, leading to non-zero concentrations of material in the gas phase. This condition is usually referred to as a vapour (Petrucci *et al* 2002). Because gas sensing applications often include extreme conditions this account will not be limited to ambient conditions, i.e. 20 °C and 1013 mbar, but all discussions are valid for atoms and molecules in the gaseous phase, including vapours. The description in this text will use the word molecule as the fundamental constituent of a gas but atoms are not excluded from this, of course. However, particulate matter is explicitly excluded since air-suspended solids or liquids are clearly distinct from gases in terms of their behaviour and therefore do not form part of the considerations presented herein.

Sensors in general, and gas sensors in particular may utilize a vast number of transducing mechanisms (Fraden 2016), which makes the field interesting from a scientific and technological point of view but also rather complex when comparing the performance of different techniques. The different properties typically give rise to cross-sensitivities as well, since each property might have an effect on the

doi:10.1088/978-0-7503-3159-3ch2

transducer itself. The possible interactions of a gas with light or matter may include collisions, electric, magnetic, electromagnetic, thermal, and chemical reactions. All of these as well as their combination may be used to construe a gas sensor and the detection of a gas species may rely on its macroscopic and/or microscopic properties. The following description is a mere introduction to the different properties and possible interactions. The depth of discussion is appropiate to understand the working principles of gas sensors. Much more profound theoretical background is available from dedicated textbooks (Jeans 1982, Chapman and Cowling 1998, Liboff 2003, Schram 2012).

2.1 Microscopic properties

On a microscopic level, molecules are made up of the fundamental constituents of matter, i.e. atoms built from protons, neutrons, and electrons (Fraser 1992). The overall arrangement of the different atoms will lend the basic properties to each molecule and the structural parameters of a molecule are mainly determined via the interaction between the electric charges of the constituent particles and the resulting charge distribution. However, a centre of mass C_{M} can be determined for each molecule, such that each molecule features a total mass m as well as electric and magnetic field distributions. A schematic drawing of a molecule is presented in figure 2.1.

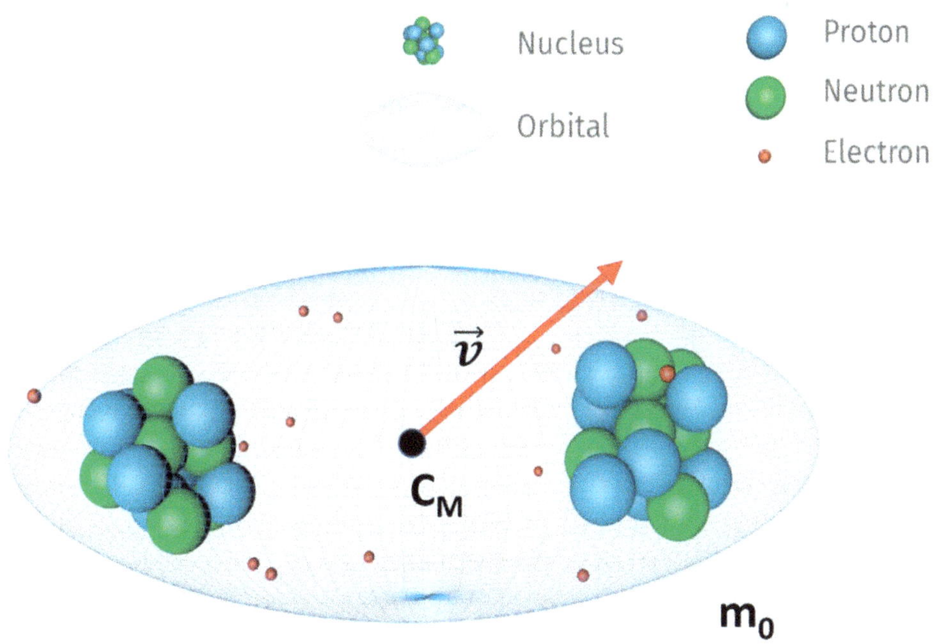

Figure 2.1. A single molecule is made up of protons (blue), neutrons (green), and electrons (red), which all contribute to the total mass of a molecule. The number of electrons in a neutral molecule equals that of its protons and their mass is negligible as compared to that of protons and neutrons. The relative position of electrons may only be described by the probability of finding them at a certain position, which is described by so-called orbitals. The centre of mass of a single molecule moves at a constant velocity v until an event changes it.

In a gas, the interaction between molecules is much less pronounced as compared to the situation encountered in liquids or solids. However, the electromagnetic interactions between molecules are responsible for a plethora of effects that enable and give rise to various methods of detection via the resulting chemical behaviour. Additionally, molecules will exhibit a movement of their centre of mass at non-zero temperatures with a certain velocity $\vec{v}$, which in turn is associated with a momentum $\vec{p} = m \cdot \vec{v}$ and a kinetic energy $E_k = \frac{m}{2}v^2$ (Tipler and Mosca 2008).

2.1.1 Mass and momentum

The number and kind of the atoms that make up a molecule determine the mass of an individual molecule. Because the mass of an eletron[1], m_e, is about 1836 smaller than the mass of protons[2], m_p, and neutrons[3], m_n, it may be disregarded in the following discussion regarding the total mass of molecules. Furthermore, since $m_p \approx m_n$ the mass of a molecule or molecular mass may be estimated by simply adding the total number of protons and neutrons. To simplify the calculus, the so-called unified atomic mass unit u or Dalton Da may be used (Taylor 2009):

$$1\ \mathrm{u} = 1\ \mathrm{Da} = 1.66{\cdot}10^{-27}\ \mathrm{kg}. \tag{2.1}$$

Hence one may estimate the molecular mass using 1 u for each proton and neutron. Because one rarely works with a single molecule, the Avogadro constant $N_A = 6.022\ 10^{23}$ is used to define the molar mass M of a molecular species:

$$M = m \cdot N_A, \tag{2.2}$$

i.e. the mass of $6.022{\cdot}10^{23}$ molecules. Looking at the numbers, it becomes clear that this system is defined in such a way as to yield a weight of approximately 1 g for $6.022{\cdot}10^{23}$ particles of mass 1 u. For example, $6.022{\cdot}10^{23}$ molecules of hydrogen (H_2) weigh about 2 g, whereas the same number of oxygen molecules (O_2), with $m_O = 16$ u for each oxygen atom, weigh approximately 32 g.

In turn this means that two molecules with equal kinetic energies but different masses will move at different speeds. While an oxygen molecule with $5 \cdot 10^{-21}$ J of kinetic energy will move at $\sim$100 m s^{-1}, a hydrogen molecule will move at $\sim$1720 m s^{-1} when possessing an equal amount of kinetic energy. Gas matrices strive to achieve thermal equilibrium, i.e. equal kinetic energy of all constituents, and not equal velocity.

2.1.2 Internal molecular states

Apart from its mass, each molecule is characterized by a number of internal energy states that are mainly governed by the distribution and movement of the electrons as well as the relative motion of the atoms in a molecule (Feynman *et al* 2011).

[1] Mass of an electron; $m_e = 9.109 \cdot 10^{-31}$ kg.
[2] Mass on a proton: $m_p = 1.673 \cdot 10^{-27}$ kg.
[3] Mass of a neutron: $m_n = 1.675 \cdot 10^{-27}$ kg.

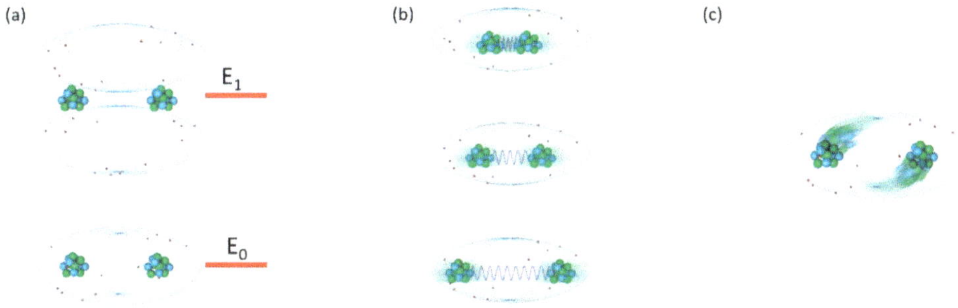

Figure 2.2. (a) The occupation of an orbital is associated with certain internal energy (here E_0 and s_1) and angular momentum levels. Additionally, each electron possesses the quantum mechanical property spin s, which also has the dimension of an angular momentum. (b) The relative movement between the atoms in a molecule takes on the form of vibrations associated with vibrational energy. (c) Rotations do carry both an energy level as well as an angular momentum.

Additionally, electronic interactions are ultimately responsible for the stability and conformation of the molecule as well as for its chemical activity. Their theoretical description requires input from quantum mechanics as well as the theory of relativity, and a number of dedicated texts are available on these topics (Helgaker 2014, Demaison and Vogt 2020). However, here only the main conclusions are presented in a nutshell. The internal energy of a molecule may be stored in different types of relative movement and setting aside processes of the nucleus, one may divide them into three main types relevant for gas sensing. Figure 2.2 schematically depicts the types of relevant internal energy levels for this book.

2.1.3 Electronic states

While the exact position of the electrons that form part of a molecule may not be pinned down, the average positions may be described by so-called orbitals (Cohen-Tannoudji *et al* 1977, Feynman *et al* 2011). They describe the probability of finding an electron at a certain position relative to the nuclei and their calculation becomes more complex the more electrons form part of a molecule (Daintith 2004). As a rule of thumb, an increase in average distance between electrons and nuclei indicates rising energy levels because separating particles that attract each other comes at a cost. This may be illustrated by looking at the orbitals of the hydrogen atom (H) as the least complex example. It is made up of a single proton and an electron. Because the electron is much lighter than the proton, the relative motion of both particles may be approximately described by the electron's movement around the proton. The strongest interaction between both is Coulomb's attraction force and the total internal energy of the hydrogen atom thus reads:

$$E_{\text{tot}} = E_{\text{kin}} + E_{\text{Cou}} = \frac{m\,v^2}{2} - \frac{e^2}{4\pi\,\varepsilon_0 r}, \tag{2.3}$$

where e is the elementary charge,[4] ε_0 the vacuum permitivity[5] and r the distance between particles. In the quantum mechanical description, the energies are expressed as operators $E_{\text{tot}} = i\,\hbar\frac{\partial}{\partial t}$, $E_{\text{kin}} = -\frac{\hbar^2}{2\,m_e}\nabla^2$, $E_{\text{Cou}} = \frac{e^2}{4\pi\,\varepsilon_0 r}$, which leads to the so-called Schrödinger equation for the hydron atom:

$$i\hbar\frac{\partial}{\partial t}\,\psi(r,\,\theta,\,\varphi) = \left(-\frac{\hbar^2}{2\,m_e}\,\nabla^2 - \frac{e^2}{4\pi\,\varepsilon_0 r}\right)\psi(r,\,\theta,\,\varphi) \tag{2.4}$$

where $\psi(r,\,\theta,\,\varphi)$ is the wavefunction in spherical coordinates describing the location of the electron. The solutions $\psi(r,\,\theta,\,\varphi)$ of this equation give rise to orbitals and describe the physical properties of a solution of the Schrödinger equation, including the energy as well as the angular momentum associated with a given orbital. On top of that, the electron possesses a spin with values $s_e = \pm\frac{\hbar}{2}$, which is a purely quantum mechanical parameter with the units of an angular momentum.

Much like in the case of the hydrogen atom, molecules and other atoms alike yield possible orbitals that the electrons may occupy and different possible orbitals possess distinct energy levels and angular momentum states associated with it. Much of the chemical activity of a molecular species is determined by the orbitals that are or are not occupied. Additionally, the magnetic behaviour of a gas is mainly governed by the orbital structure. Notably, gases may exhibit paramagnetic or diamagnetic behaviour depending on the electronic configuration. The internal energy that may be stored in electronic states is typically on the order of $\sim$1 eV per molecule or above, giving it the largest internal energy potential.

2.1.4 Vibrational states

Since a molecule is made up of at least two atoms it may exhibit relative movement between the various nuclei. That means that apart from the centre of mass movement the constituents of a molecule may move with respect to each other. One possible form of movement is vibrations, i.e. atoms oscillating periodically with respect to each other (Wilson *et al* 2019). The possible movements are conditioned by the structure of the molecule, which in turn is determined by structure of the orbitals. Again, the possible forms of vibrations as well as the associated vibrational frequencies are mainly determined by the rules of quantum mechanics, and theoretical solutions may be obtained by analyzing the corresponding Schrödinger equation. In a very coarse approximation molecular vibrations may be described by a harmonic oscillator, i.e. the restoring force F_{HO} is proportional by a factor k to the displacement x from the equilibrium position of the molecule's nuclei x_0 yielding a potential energy (Tipler and Mosca 2008):

$$V_{\text{HO}}(x) = \frac{k}{2}\cdot x^2 = \frac{m\cdot\omega^2}{2}\cdot x^2. \tag{2.5}$$

[4] $e \approx 1.602\cdot 10^{-19}$ C.
[5] $\varepsilon_0 \approx 8.854\cdot 10^{-12}\,\frac{F}{m}$.

The quantum mechanical solution to this problem results in discrete harmonic oscillator levels of the form (Blaise 2011):

$$E_n = \hbar \cdot \omega \cdot \left(n + \frac{1}{2}\right), \tag{2.6}$$

where n is a natural number, meaning that only discrete vibrational frequencies are possible. However, the harmonic approximation for molecular vibrations breaks down very quickly, which is why the so-called Morse potential is introduced to describe the potential energy of molecular vibrations (Costa Filho *et al* 2013):

$$V_{\mathrm{Morse}}(x) = D_e \cdot (1 - e^{-a(x-x_0)})^2, \tag{2.7}$$

wherein D_e and a describe the depth and width of the potential, respectively. The vibrational energy levels for molecules now read:

$$E_\upsilon = \hbar \cdot \omega \cdot \left(\upsilon + \frac{1}{2}\right) - \frac{(\hbar\,\omega)^2}{4\,D_e} \cdot \left(\upsilon + \frac{1}{2}\right)^2, \tag{2.8}$$

where υ is a non-negative integer. The energy stored in a vibrational state of a molecule is typically on the order 100 meV per molecule. The total number of vibrational modes of a molecule scales with the number of its constituents. While N independent atoms possess $3N$ degrees of translational freedom, a molecule made up of N atoms only features three translational degrees of freedom, namely those of its centre of mass. A further three degrees of freedom are covered by molecular rotation, which leaves the total number of vibrational modes at $3N - 6$. The exception to this rule is linear molecules, since they lack rotational mode, which in turn leads to a total of $3N - 5$ vibrational modes.

2.1.5 Rotational states

Lastly, a molecule might rotate around its centre of mass. The associated moment of intertia in the classical description reads (Tipler and Mosca 2008):

$$I = \sum_i m_i r_i^2, \tag{2.9}$$

where m_i is the mass of the respective nuclei and r_i its distance to the centre of mass. The associated angular momentum L for a rotational frequency ω is:

$$L = I \cdot \omega. \tag{2.10}$$

While a quantum mechanical description applicable to all types of molecules is rather complex, the basic lessons may be learned by limiting the discussion to a two-atomic gas with a constant distance between both nuclei, i.e. a rigid rotator. To comply with the restrains of quantum mechanics, the angular momentum has to be a multiple of $\hbar$ and the energy associated with rotation may be expressed as (Baer and Hase 1996):

$$E_J = \frac{\hbar^2}{2I} \cdot J \cdot (J + 1) = B \cdot J \cdot (J + 1), \tag{2.11}$$

where $B = \frac{h}{2 \cdot \pi^2 \cdot c \cdot I}$ is the rotational constant that takes into account the moment of intertia of the molecule I, the structure and masses involved in the formation of the molecule. It is important to note here that a rotational state J with energy E_J has a total of $2J + 1$ different, so-called degenerate sub-states that all possess the same energy. The difference in energy between two neighbouring quantum numbers J reads:

$$E_{J+1} - E_J = B \cdot (J + 1) \cdot (J + 2) - B \cdot J \cdot (J + 1) = 2 \cdot B \cdot J + 2, \qquad (2.12)$$

The energy that is stored in rotation is typically on the order of 10 meV. Apart from that energy, the rotation does possess an angular momentum.

2.2 Macroscopic properties

In a first approximation, the macroscopic properties of a gas may be described by the kinetic theory of gases (Schram 2012). This simple, classical theory combines the thermodynamic, macroscopic properties of a gas, i.e. temperature, pressure, and volume. To derive expressions relating these properties, a few simple assumptions are made, namely that:

- the number of gas molecules is large such that the thermodynamic limit applies;
- the separation between any two molecules is on average much bigger than their size;
- collisions between the container and molecules as well as inter-molecular collisions are elastic;
- no interactions other than collisions take place;
- the movement of molecules is random.

It is important to keep in mind that once one or more of those assumptions fail to adequately describe a gas, the rules of behaviour deduced from those assumptions are likely to no longer hold. One important example is the influence of internally available degrees of freedom of a molecule that will influence the macroscopic behaviour, especially the transport properties. This is why the microscopic structure of gases does influence the possibly suitable transducing mechanisms.

Based on Newton's laws of classical mechanics, microscopic properties of individual molecules may now be connected with macroscopic properties of a gas and at this stage the internal structure of the molecules is not considered. Figure 2.3 illustrates the behaviour of an ideal classical gas in a container including the interactions that may arise, namely collisions with the walls of the container and between two molecules. In thermodynamic equilibrium, the kinetic energy of different gas species is equal, which in turn means different velocities for different molecular masses.

2.2.1 Temperature

The random movement of particles is associated with the non-directional part of the kinetic energy E_{kin} of an ensemble of molecules. Since the notion of temperature is associated only with the random movement of molecules, kinetic energy refers only to that part of the total kinetic energy that is associated with random movement, i.e.

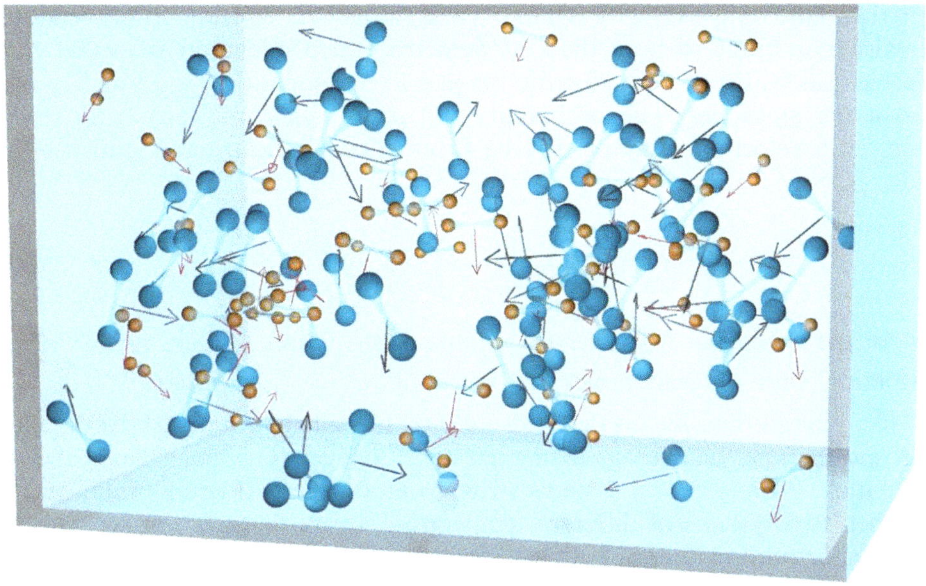

Figure 2.3. Molecules in a container move in a random direction and with a velocity distribution that is determined by the temperature of the gas. The macroscopic properties are a result of the microscopic properties.

excluding any overall directed centre of mass movement that an ensemble might have. Hence each individual molecule possesses a kinetic energy E_{kin} as a function of its velocity $\vec{v}$ and mass m (Nolting 2016):

$$E_{\text{kin}} = \frac{m\,v^2}{2}, \tag{2.13}$$

The thermal energy $E_{\text{kin,ens}}$ of an ensemble of N molecules is consequently (Kittel and Krömer 1980):

$$E_{\text{kin,ens}} = N \cdot \overline{E_{\text{kin}}} = N \cdot \frac{m\,\overline{v^2}}{2}, \tag{2.14}$$

i.e. N times the average kinetic energy of a molecule. The concept of temperature may be understood as a measure of the thermal energy of an ensemble of molecules and may be written as:

$$T = \frac{m\,\overline{v^2}}{3 \cdot k_{\text{B}}}, \tag{2.15}$$

where k_{B} is the so-called Boltzmann constant[6] and T the absolute temperature in Kelvin (K). In consequence, this means that in this simple model the temperature is an absolute measure and may not be negative. If the molecules of a gas do not move

[6] Boltzmann constant: $k_{\text{B}} = 1.38\ 10^{-23}\ \text{J K}^{-1}$.

at all, then the average velocity is 0 m s^{-1} and the associated temperature reads 0 K. It is also important to note that the velocity is dependent on the mass of the molecule. Lastly, the velocity distribution at a fixed temperature is a function of the mass of the molecules. The probability $f(v)$ of finding a molecule with mass m moving at a velocity v may be derived from Maxwell–Boltzmann statistics to be (Müller-Kirsten 2013):

$$f(v) = \left(\frac{m}{2\pi \cdot k_B \cdot T}\right)^{\frac{3}{2}} \cdot e^{-\frac{mv^2}{2k_B T}}. \quad (2.16)$$

This is the so-called Maxwell–Boltzmann distribution, which is a Gaussian distribution with a mean molecular speed of $\langle v \rangle = \frac{2}{\sqrt{\pi}} v_p$ with $v_p = \sqrt{\frac{2\,k\,T}{m}}$ the most likely speed. This means that a gas of, say, 10^9 pure hydrogen (H_2) molecules at 300 K moves faster on average than a gas of 10^9 oxygen (O_2) molecules at the same temperature. Also, physical systems strive to attain thermal equilibrium, meaning that when two systems of different temperature are in thermal contact with each other, the thermal energies of both will change over time until both have reached the same temperature, not the same velocity distribution.

2.2.2 Pressure

Another concept describing a macroscopic quantity of a gas that is associated with its velocity and mass is the so-called pressure p. It may be interpreted as a measure of the number of collisions the gas molecules have with the container walls per unit time. Each collision results in a force onto the wall with an area A. Pressure is defined as the average force $\overline{F}$ of the ensemble of molecules onto the walls:

$$p = \frac{\overline{F}}{A}. \quad (2.17)$$

Assuming that the container is a cube with edges of length l the volume V of the gas is $V = l^3$ and the wall area $A = l^2$, which results in a pressure of a gas made up of N molecules:

$$p = \frac{\overline{F}}{l^2} = \frac{N \cdot m\,\overline{v^2}}{3\,l^3} = \frac{N \cdot m\,\overline{v^2}}{3\,V}. \quad (2.18)$$

The mean free path length λ of a molecule is the path it can travel on average before it collides with another molecule. It is given by the mean velocity of the molecule $\langle v \rangle$ and the inverse frequency of collisions, which is proportional to the pressure p.

Comparing equations (2.15) and (2.18) leads to the so-called ideal gas law, which relates pressure, volume, and temperature:

$$p \cdot V = N \cdot k_B \cdot T. \quad (2.19)$$

A more realistic description of real gases is obtained by performing a Taylor expansion of the ideal gas law and using this to correct for, e.g., effects of interaction between molecules.

In this regard, it is important to note and keep in mind that the internal molecular properties do influence the macroscopic properties to a considerable degree. In particular, the structure of internal molecular states has a sizeable influence on macroscopic behaviour.

2.3 Transport properties

A gaseous medium is also able to transport physical quantities from one location to another. In particular, this includes mass, kinetic energy, and momentum. The overall amount of each of these quantities is conserved and given any system's impetus for equilibrium conditions, any gradient within a gases' container will lead to a flux in the respective quantity. In a generic sense, flux is the flow of a quantity through an area and this is an adequate model to describe the transport processes. This means that any gradient in any of those physical quantities will trigger a transport mechanism, which might be, e.g., the molecules themselves, their thermal energy, or their momentum. The concept is schematically depicted in figure 2.4.

To exemplify this: If there are different temperatures in a lecture hall, then heat will flow from the warmer region to the colder region until thermal equilibrium is achieved and all places possess the same temperature. In general, any gradient will lead to a flux of the respective quantity with the ultimate goal to achieve equilibrium throughout the container in that quantity. By using the properties from the short introduction into the macroscopic behaviour of gases, some of those transport properties may be derived.

2.3.1 Kinetic energy transport—thermal conductivity

The interaction via a thermal contact between regions of different temperatures will lead to processes aiming for thermal equilibrium. Assuming no concentration difference of gas molecules, the collisions between molecules will ultimately transfer kinetic energy from a region of higher temperature to one of lower temperature. The functional relation between heat flux $\vec{Q}$ and temperature gradient ΔT is a linear relation:

$$\vec{Q}(\vec{r}, t) = \kappa \cdot \Delta \mathrm{T}(\vec{r}, t), \tag{2.20}$$

where κ is the thermal conductivity of the gas species. Importantly, the transfer of heat via thermal conduction does not require any transfer of mass, i.e. the random

Figure 2.4. Transport processes are typically the result of non-equilibrium conditions of physical quantities. Here the diffusion of red particles with the lapse of time is schematically depicted. This transport of mass visualizes the flux of a physical quantity to obtain an equilibrium state, i.e. a homogenous number density of the constituents of the mixture.

movement of molecules is not superimposed by an overall mass flow from one region to another. Also note that the internal degrees of freedom as well as the molecular mass will influence the thermal conductivity of a gas.

2.3.2 Mass transport—diffusion

A difference in the mean molecular density of a gas species in a container will lead to a mass flux in order to compensate the existing gradient and attain equal concentration of all as species throughout the container. The flux of molecules is a function of the so-called diffusion coefficient D, and the concentration gradient $\nabla\varphi$:

$$\vec{J}(\vec{r}, t) = -D \cdot \nabla\, \varphi(\vec{r}, t). \tag{2.21}$$

This simple relation is better known as Fick's first law (Fick 1855) and it describes the flux of molecules. The value of the diffusion coefficient depends on the temperature .via its relation to the mean velocity of the molecules and on the pressure via the mean free path length in the case of an ideal gas. In actual scenarios an expression for D may be obtained by considering the scattering probabilities between different species of molecules.

Since diffusion changes the concentration gradient over time, it is clear that both the flux and the concentration are time dependent, and the resulting differential equation is known as Fick's second law:

$$\frac{\partial\varphi(t)}{\partial t} = D \cdot \Delta\varphi(t). \tag{2.22}$$

The solution of this differential equation is an exponential function in time, which means that any concentration gradients will diminish exponentially with time and as a function of the diffusion coefficient. The properties of diffusion are important when considering the performance of sensing apparatus but as a means to implement a transducer, diffusion plays a secondary role.

2.3.3 Momentum transport—viscosity

Lastly, the so-called viscosity quantifies how a gas reacts if layers of the gas move with respect to each other. This effectively results in a sheer stress and it is proportional to the velocity gradient between both sheets. This ultimately leads to an equilibrium situation, where both sheets are equally fast. This is equal to the transport of momentum from one sheet to another and the proportionality factor is called viscosity μ. For the molecules of a gas the viscosity may be estimated by:

$$\mu = \frac{1}{3} \cdot n \cdot m \cdot \langle v \rangle \cdot \lambda, \tag{2.23}$$

where n is the number density. For gas sensing transducer mechanisms the viscosity is often of secondary importance.

References

Daintith J (ed) 2004 *A Dictionary of Chemistry* 5th edn (Oxford: Oxford University Press) (Oxford reference online premium)

Baer T and Hase W L 1996 Vibrational/rotational energy levels *Unimolecular Reaction Dynamics: Theory and Experiments* (The International Series of Monographs on Chemistry, 31) T Baer and W L Hase (Hg.) (New York: Oxford University Press)

Blaise P 2011 *Quantum Oscillators* (Hoboken, NJ: Wiley)

Chapman S and Cowling T G 1998 *The Mathematical Theory of Non-Uniform Gases. An Account of the Kinetic Theory of Viscosity, Thermal Conduction and Diffusion in Gases* (Cambridge Mathematical Library) 3rd edn Transferred to digital printing (Cambridge: Cambridge University Press)

Cohen-Tannoudji C, Diu B and Laloë F 1977 *Quantum Mechanics* (Textbook Physics, 1) (New York: Wiley)

Costa Filho R N, Alencar G, Skagerstam B-S and Andrade J S 2013 Morse potential derived from first principles *EPL* **101** 10009

Demaison J and Vogt N 2020 *Accurate Structure Determination of Free Molecules* (Lecture Notes in Chemistry, 105) 1st edn (Cham: Springer International Publishing)

Feynman R P, Leighton R B and Sands M L 2011 *The Feynman Lectures on Physics* 2nd edn (San Francisco, CA: Addison-Wesley)

Fick A 1855 V. On liquid diffusion *Lond. Edinb. Dubl. Phil. Mag. J. Sci.* **10** 30–9

Fraden J 2016 *Handbook of Modern Sensors. Physics, Designs, and Applications* (Lecture Notes in Chemistry, 105) 5th edn (Cham: Springer International Publishing)

Fraser G (ed) 1992 *The New Physics* (Cambridge: Cambridge University Press)

Goodstein D L 2014 *States of Matter* (Dover Books on Physics) (New York: Dover)

Helgaker T 2014 *Molecular Electronic-Structure Theory* (Chichester: Wiley)

Jeans J H 1982 *An Introduction to the Kinetic Theory of Gases* (Cambridge Science Classics) (Cambridge: Cambridge University Press)

Kittel C and Krömer H 1980 *Thermal Physics* 2nd edn (San Francisco, CA: Freemann)

Liboff R L 2003 *Kinetic Theory. Classical, Quantum, and Relativistic Descriptions* 3rd edn (Graduate Texts in Contemporary Physics) (New York: Springer)

Müller-Kirsten H J W 2013 *Basics of Statistical Physics* 2nd edn (Singapore: World Scientific)

Nolting W 2016 *Classical Mechanics* (Theoretical Physics) (Cham: Springer)

Petrucci R H, Harwood W S and Herring F G 2002 *Química General: Principles and Modern Applications* 8th edn (Madrid: Prentice-Hall)

Schram P P J M 2012 *Kinetic Theory of Gases and Plasmas* (An International Book Series on the Fundamental Theories of Physics, 46) (Dordrecht: Springer)

Taylor B N 2009 Molar mass and related quantities in the New SI *Metrologia* **46** L16–9

Tipler P A and Mosca G 2008 *Physics for Scientists and Engineers* 6th edn (New York: W.H. Freeman)

Wilson E B, Decius J C and Cross P C 2019 *Molecular Vibrations: The Theory of Infrared and Raman Vibrational Spectra* (Dover Books on Physics) (New York: Dover)

Chapter 3

Direct transducers for gas sensing

The wide range of requirements for a plethora of scenarios requiring gas sensors has led to many different gas detection technologies existing in parallel. The parameter space for gas sensing applications is constructed by at least sensitivity, selectivity, and stability to which size, robustness, and price are usually added. To evaluate the distinct capabilities and potentials of different approaches, this chapter gives a short introduction to the working principle of direct transducers, i.e. those technologies that directly convert chemical information into an electrical signal. It intends to organize the different techniques by their degree of selectivity and provide a rough estimation with regards to the respective potentials for miniaturisation.

Gas sensing is historically rooted in the mining business and it has been around far longer than any of the technologies used at present. Current applications of gas sensors are still centred around safety and security scenarios but analytical and consumer demands have increased considerably in recent years and are likely to do so even more in the future. From a strictly technological point of view it is today possible to detect and identify almost any arbitrary molecular species and determine its quantity down to a single molecule level. However, size, cost and infrastructure may prevent the deployment of such specific and sensitive equipment. That is why it is important to keep in mind the requirements for the use of any type of sensor in a specific application. Any user has to be clear about what is needed for their measurement task and to what extent selectivity, stability, size, and sensitivity matter. This is why many different techniques are currently deployed and co-exist. At this moment there simply is no one-size-fits-all solution for chemical sensing and this makes the understanding of the capabilities of the distinct approaches important.

Consequently, prior to choosing a technology one has to clarify if a selective or even specific detection of one or more gas species is necessary. Secondly, it is paramount to know if a quantitative determination is required and what the performance in terms of limit of detection, dynamic range, and resolution needs

doi:10.1088/978-0-7503-3159-3ch3

to be. Thirdly, the conditions of measurement have to be identified. The number of possible parameters influencing the sensor output depends on the transducing mechanism, which is why the basic working principles are briefly discussed first. After all, the aim is to build a sensor, which in this case means the conversion of chemical information into an electronic signal. The transducing principles presented in this chapter are able to directly convert chemical information into an electronic signal and the principles governing their behaviour and the transducing mechanism are explained.

All sensors do exhibit unintended cross-sensitivities to other parameters and in gas sensing this explicitly includes other gas species apart from further magnitudes such as temperature, pressure, or electromagnetic fields. However, a coarse classification or ordering of the degree of selectivity is attempted within this chapter.

The basic transducing mechanism discussed often forms part of more complex gas analyzers, which will be discussed in chapter 5. In this regard, the sensitivity and selectivity towards various types of gases as well as the potential limits of detection are part of the discussion.

3.1 Thermal conductivity detectors

Physical systems strive to attain thermal equilibrium if no constraints are placed upon them and thermal conductivity discussed in the previous chapter is one manifestation of this general behaviour (Saggion *et al* 2019). This means that components will converge into a state of equal temperature eventually. To achieve this, three basic processes are available to transfer heat from one place to another (Böckh and Wetzel 2012):

1. Radiation.
2. Convection.
3. Conduction.

The latter has been introduced already but in order to evaluate a system's behaviour all three processes should be considered. A schematic depiction of the three processes is shown in figure 3.1.

3.1.1 Radiation

The emission of radiation by matter is a fundamental property of all matter at temperatures $T > 0$ K. Since electromagnetic radiation carries energy and the total energy is conserved, this means matter loses energy via radiation hence this is a heat loss process. It is a basic property of all objects and the idealized model to describe this behaviour is via so-called black body radiation (Stewart and Johnson 2017). Establishing an expression for the spectral distribution of a black body has been an important cornerstone in the development of quantum theory and is closely connected to the introduction of Planck's constant[1], h. According to

[1] $h = 6.626 \cdot 10^{-34}$ J·s has been introduced by Max Planck as the so-called 'Hilfskonstante', i.e. help constant.

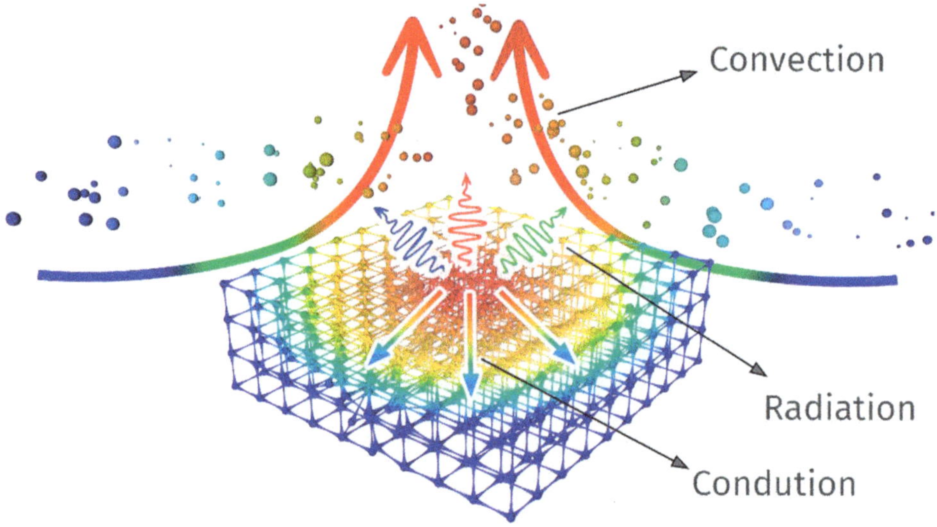

Figure 3.1. Schematic illustration of heat loss mechanisms. While radiation and conduction do not involve the transfer of mass, convection is associated with transporting molecules of higher temperature away from the heated region.

Planck's law the spectral radiance[2] $B_\nu(T)$ of a body with temperature T is given by (Planck 1989):

$$B_\nu(T) = \frac{2 \cdot h}{c_0^2} \frac{\upsilon^3}{e^{\frac{h\nu}{k_B T}} - 1}, \tag{3.1}$$

where ν and c_0 are the radiation frequency and speed of light, respectively. This means that the spectral distribution of electromagnetic waves emitted from a black body is determined only by its temperature. All other parameters in equation (3.1) are physical constants. Interestingly, describing the real behaviour of actual matter at a finite temperature does not require fundamental changes to this expression. By scaling the formula of black body radiation by the emissivity ε_ν, which is a property of the material and has values between 0 and 1, it is possible to describe actual, so-called grey body radiation. Analyzing Planck's law allows for assessing the general behaviour of matter as a function of temperature.

(a) Integrating equation (3.1) over all frequencies and all directions that radiation may be emitted to, allows for calculating the total power lost via radiation P_{rad} by a body with a surface area A and temperature T (Boltzmann 1884):

$$P_{\text{rad}} = \frac{2\pi^5 k_B^4}{15 h^3 c^2} \cdot A \cdot T^4 = \sigma \cdot A \cdot T^4. \tag{3.2}$$

[2] The unit of spectral radiance is $\text{W}\cdot\text{sr}^{-1}\cdot\text{m}^{-2}\cdot\text{Hz}^{-1}$.

This describes the so-called Stefan–Boltzmann law with $\sigma \approx 5.67 \cdot 10^{-8}$ W m^{-2} K^{-4} being a constant of the same name yielding the total electromagnetic power emitted from a surface of temperature T. This means that heat loss via radiative routes scales as T^4 and is proportional to the surface of the object. It is worthwhile to stress that this loss of energy is due only to the surface and dependent on the surface's temperature. The bulk of an object does not play a role in black body radiation.

(b) The maximum of Planck's black body curve indicates in which spectral interval the most power is irradiated. It may be calculated setting the first derivate of Planck's law to 0, yielding (Mehra and Rechenberg 1982):

$$\upsilon_{\max} \approx 5.88 \cdot 10^{10}\,\mathrm{Hz}\,\frac{T}{K} \quad \text{or} \quad \lambda_{\max} \approx 2897.8\mu\mathrm{m}\,\frac{1}{T/K}. \tag{3.3}$$

The relation is known as Wien's displacement law and allows for calculating the peak emission. It is important to note that no two Planck curves for a body ever intersect for any temperature at any frequency. This means emission at a pre-defined frequency will always be higher for higher temperatures. However, the total power irradiated, the ratio of spectral irradiance at different frequencies, as well as the maximum of the Planck curve do change with temperature.

3.1.2 Conduction

The thermal conductivity κ of matter has been introduced in chapter 2 and it is governed by different mechanisms depending on temperature, type, and state of matter. To this end, the focus here is placed on the behaviour of the thermal conductivity of gases. Like radiation, thermal conduction does not lead to a transport of mass but kinetic energy is passed on via collisions, regardless of the state of matter. For gases thermal conductivity is a characteristic magnitude that may be described theoretically to a good degree by considering the internal energy and geometric structure of molecules as well as pressure and temperature. The Chapman–Ensko theory provides means to express the thermal conductivity approximately using the gases' viscosity μ and their mass m for various models of intermolecular interaction (Chapman and Cowling 1998):

$$\kappa = f \cdot \mu \cdot \frac{k_B}{m}, \tag{3.4}$$

where f is a numberical factor close to 2.5. With respect to gas sensing and the potential for selectivity based on thermal conductivity, the interpretation of this formula yields the important conclusion that mass is a decisive factor. Namely, hydrogen and helium stand out from all other gases because they are several times lighter than any other gas.

3.1.3 Convection

Any object embedded in a fluid loses heat via transferring energy to the constituents of that fluid by means of conduction. In consequence, this leads to an increase in local kinetic energy of the fluid, which in turn leads to a volume flow of hot fluid. This means that mass is transported and along with that mass heat is transferred. The heat loss from a surface with area A and temperature T to a fluid with temperature T_f may be described as (Zhao 2022):

$$P_C = h \cdot A \cdot (T - T_f)^b, \tag{3.5}$$

where h is the heat transfer coefficient and b a constant depending on the material systems involved in the process. Importantly, convection is a complex process and involves conduction as a first step.

For the understanding of the setup of thermal conductivity gas sensors and evaluating the possible cross-sensitivities the distinct heat loss mechanisms are crucial. While precise theoretical models and simulations are a complex task, the scaling of the different mechanisms with temperature holds.

3.1.4 Basic sensor setup

As the name suggests, thermal conductivity sensors rely on the determination of the difference in the thermal conductivity of gas species (Poole 2005, Daynes 1933). The basic working principle relies on heating up an object and determining how much heat is lost as a function of the gas composition (Gardner *et al* 2023). Equation (2.20) is therefore central to the functioning of this type of sensor, i.e. the heat flux $\vec{Q}$ at a constant temperature difference will depend on the thermal conductivity κ of the gas mixture. Text for further reading on the basics and state of the art in thermal conductivity sensors are presented in table 3.1.

To make this work properly it is important to design the sensor such that radiation and convection do not influence the results or at least only to a well-defined level. Hence building a gas sensor based on this concept requires turning off the influence of both, or at least being able to account for both effects, such that the sensor only measures the thermal conductivity of the gas mixture. A schematic setup of thermal conductivity sensors is depicted in figure 3.2 and comprises two equally constructed devices operating at the same temperature. To prevent active cooling of the heating elements, a diffusion barrier is used such that no external air flow shall occur inside the measurement chambers. This means that effects of radiation loss on the sensor signal are cancelled by operating both channels at equal temperature and that convection is prevented via constructive means. The reference channel is sealed

Table 3.1. Possible starting points for further reading on sensors based on thermal conducity of gas samples.

Authors	Title	References
Gardner *et al*	'Micromachined Thermal Gas Sensors—A Review'	Gardner *et al* (2023)
Tsederberg	'Thermal Conductivity of Gases and Liquids'	Thodos (1965)
Hiller and Baldwin	'Thermal Conductivity Detectors'	Hiller and Baldwin (2005)

off from the environment in order to maintain a constant gas composition. The measurement channel allows for gas exchange with the environment.

This way, radiative losses as well as convection, which are considerably less sensisitive to gas matrix composition, are approximately equal, leaving only the thermal conductivity of the gas sample as a variable. A change in thermal conductivity due to a change in gas composition then leads to a change in temperature of the heating element. Usually, the heater is made of a metal, whose electrical resistivity is dependent on temperature. That way, the heating element itself acts as a temperature sensor as well. The read-out circuit for this type of sensor is usually a Wheatstone bridge, since it is able to detect small differences in the resistances it compares.

As an alternative operational mode, the temperature of both heating elements may be actively controlled and the power consumption necessary to maintain the temperature is utilized as sensing signal. Alternative or simpler sensor designs are possible but they will have to take into account the effects of radiation and convection on the resulting signal.

3.1.5 Selectivity evaluation and miniaturization

The level of selectivity of this type of sensor is limited since any change in thermal conductivity of the gas matrix will lead to a signal. In general, the thermal conductivity of a gas varies with its molar mass and this makes two gases stand out from the rest: helium and hydrogen. Table 3.2 presents an overview of the thermal conductivity values of various gases as a function of their respective molar masses.

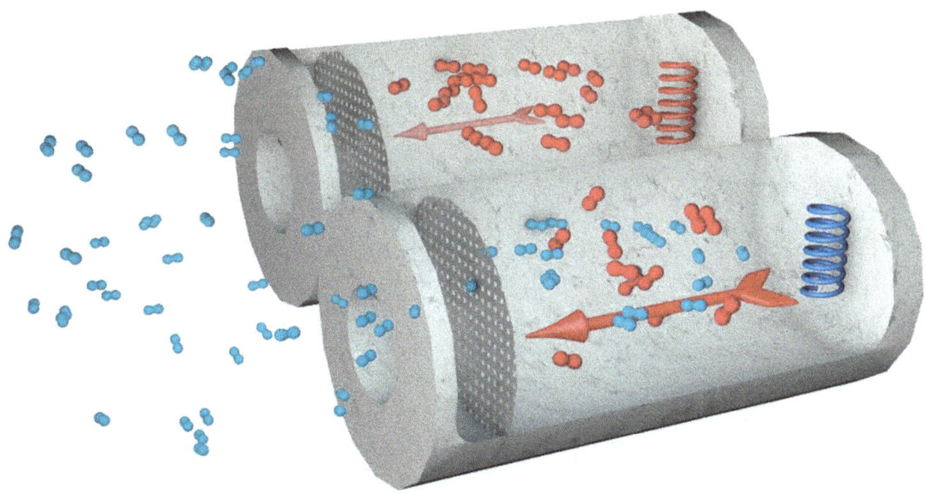

Figure 3.2. Thermal conductivity gas sensors make use of the difference in thermal conductivity of gases. A better thermal conductivity will lead to a more pronounced cooling of the heater. By using a reference and a measurement channel the sensing setups are able to cancel out the effects of alternative heat losses via radiation and convection. This in turn leads to κ being the only variable in the system's setup.

Table 3.2. Thermal conductivity value of various gas species at 300 K temperature. The values vary with temperature, which has to be factored in when analyzing the signal.

Gas species	Thermal conductivity ($mW \cdot m^{-1} \cdot K^{-1}$)	References
Dry air	26.4	Lemmon and Jacobsen (2004)
Hydrogen (H_2)	186.6	Mehl *et al* (2010)
Nitrogen (N_2)	26.0	Lemmon and Jacobsen (2004)
Oxygen (O_2)	26.5	Lemmon and Jacobsen (2004)
Carbon dioxide (CO_2)	25.2	Vesovic *et al* (1990)
Methane (CH_4)	34.4	Hellmann *et al* (2009)
Propane (C_3H_8)	18.5	Marsh *et al* (2002)

The underlying data are taken from the cited reference and highlight that equation (3.4) is a decent description of actual values of κ observed, i.e. the molecular mass plays a vital role for thermal conductivity values. It is also important to note that thermal conductivity values do depend on the temperature, which has to be considered for the different measurement scenarios. The detection of hydrogen and helium is possible with a fairly high degree of selectivity because their thermal conductivity is almost one order of magnitude higher. However, a quantitative analysis of arbitrary gas matrices is difficult with this type of gas sensor. Nonetheless, it is a powerful tool for the analysis of binary gas mixtures and this is in fact the main application scenario for thermal conductivity sensors, i.e. measurement tasks where only two components are presented to the sensor at once.

The miniaturization of this sensor concept by means of microsystems technology is very much possible and an ongoing effort. The integration of compensation methods for radiation and convection are among the main design issues.

3.1.6 Application scenarios

Whenever a gas composition is mainly or entirely made up of only two types of gases and the relative concentration of those two components is sought, thermal conductivity sensors are good candidates. The sensitivity of thermal conductivity sensors depends on the operational parameters but may achieve limits of detection in the part-per-billion (pbb) range under laboratory conditions. Two examples for applications include the use in biogas quality assessment and gas chromatography.

3.1.6.1 Biogas

The fermentation of biomass has been used for the production of methane (CH_4) for many years. The process mainly produces methane (CH_4) and carbon dioxide (CO_2), which jointly typically account for >95% of the gas mixture leaving the fermenter. Other gas species include nitrogen (N_2), oxygen (O_2), hydrogen sulphide (H_2S), carbon monoxide (CO), and water vapour (H_2O). The latter is usually saturated, meaning that its content may be considered constant at constant temperature, while all other gases usually occuring only as trace gases, i.e. with concentrations below 1%. To determine

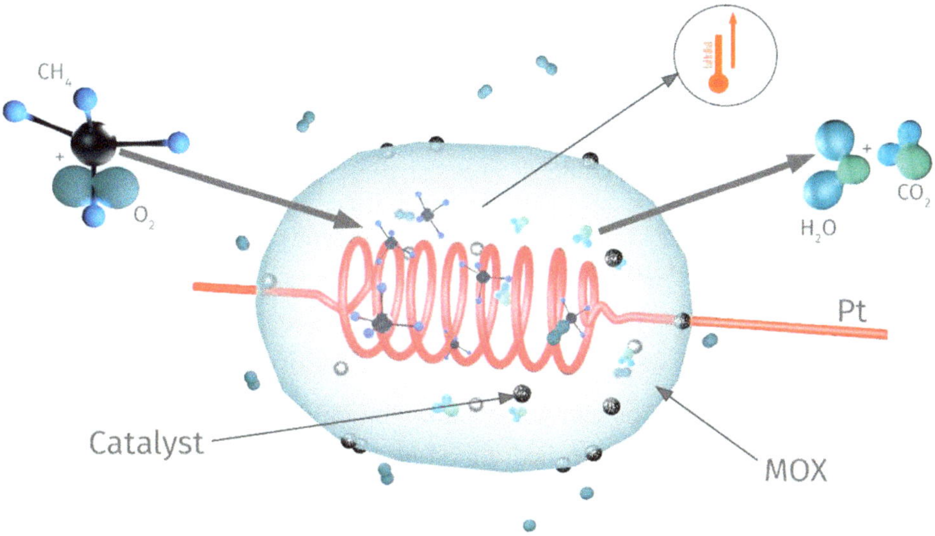

Figure 3.3. Schematic drawing of a pellet resistor featuring a catalyst decorated metal oxide (MOX) bead heated by a platinum (Pt) wire: The exothermic reaction of molecules in the vicinity of the heating wire causes an increase in temperature, which in turn is detected via a change in resistivity.

the quality of biogas, the methane content is the main indicator and thermal conductivity detectors may be used to deliver this information by assuming that at room temperature $\kappa_{CO4} = 34.1$ mW m^{-1} K^{-1} and $\kappa_{CO2} = 16.8$mW m^{-1} K^{-1}, i.e. a ratio of about 2.03. Obviously, the technique is prone to cross-sensitivities from other trace gases in biogas but for many applications the current precision of available systems of 1% is sufficient.

3.1.6.2 Gas chromatography

A standard application for thermal conductivity sensors is gas chromatography, which is designed to produce binary mixtures of a carrier gas and a second component. The setup is slightly different from the one depicted in figure 3.3, in that both the reference and measurement channel are exposed to a constant and equal gas flow. The measurement channel contains both the carrier gas as well as the gas sample, and any deviation form a pure carrier gas will lead to a signal. If the thermal conductivity value of the second component is known, then the sensor output may be calibrated, such that a quantitative analysis of that component is possible.

3.2 Paramagnetic sensors

The magnetic properties of gases are a result of the electronic configuration and the vast majority of gas species exhibit diamagnetic behaviour, i.e. they are repelled by inhomogeneous magnetic fields (Jasek *et al* 2022). In other words, the magnetization of the gas $\vec{M}$ as a result of an applied magnetic field strength $\vec{H}$ is of opposite sign:

$$\vec{M} = \chi_m \cdot \vec{H}, \tag{3.6}$$

which means that the magnetic susceptibility χ_m is negative. However, oxygen (O_2) and mononitrogen oxides (NO and NO_2) are paramagnetic, i.e. their magnetic susceptibility χ_m is possive. This in turn enables the construction of gas sensors that make use of this property to build devices for those gases with a positive χ_m. The magnetic susceptibility of several gases is summarized in table 3.3.

According to Curie's law (Carlin and van Duyneveldt 1977), the magnetic susceptibility is temperature dependent:

$$\chi_m = \frac{C}{T}, \tag{3.7}$$

where $C = \mu_0 \cdot n \cdot \frac{\mu^2}{3k_B}$ is the Curie constant. The utilization of this dependency is that the magnetization may be turned off via increasing the temperature of the gas.

The working principle of paramagnetic gas sensors relies on leveraging the paramagnetic properties of a gas and features two main steps (Pauling *et al* 1946):

1. The paramagnetic property means that an inhomogeneous magnetic field may be used to attract this type of gas, since they will follow the magnetic field strength. Thus, by applying an inhomogeneous magnetic field, a paramagnetic gas species may be attracted. Oxygen is a natural candidate, since it is highly paramagnetic. This attractive force leads to a molecular flow of only the paramagnetic species, which is superimposed on the flow of the gas composition as a whole.
2. The resulting additional gas flow is detected by a secondary transducing mechanism. The strength of that flow is dependent on the concentration of the paramagnetic components only and the detection of that additional flow may be achieved by several techniques. The different detection strategies typically rely on induced changes of thermal or mechanical properties:
    1. (A) Thermo-magnetic read-out

        An inhomogeneous magnetic field is used to direct paramagnetic molecules towards a heating element (Dyer 1947). Once they reach the heater, the magnetization is effectively switched off according to equation (3.7), such that the attractive force is no longer active for those heated molecules. The

Table 3.3. Magnetic susceptibility values of various gas species at 300 K temperature.

Gas species	Magnetic susceptibility/$\chi_m \cdot 10^{10}\mathbf{m^3/mol}$	References
Nitrogen (N_2)	−1.51	Rumble (2022)
Oxygen (O_2)	429.1	Rumble (2022)
Carbon dioxide (CO_2)	−2.64	Rumble (2022)
Nitrogen dioxide (NO_2)	18.1	Rumble (2022)
Mononitrogen oxide (NO)	183.6	Rumble (2022)

continuous flow of cold molecules pushes those demagnetized molecules away and effectively results in a gas flow that only exists because of paramagnetic gas species in the gas composition. This means that the attraction of paramagnetic molecules leads to an additional flow that will cool the heating element and by determining the actual temperature of the heater the amount of paramagnetic gas may be determined. The transducing mechanism is the same as in thermal conductivity sensors but now determining convection. There are various construction possibilities for this sensor type but the underlying principle is the same. It is to be clearly distinguished from thermal conductivity sensors in that here a gas flow of paramagnetic molecules is generated via an inhomogeneous magnetic field and the cooling effect of that flow on the heating element's temperature is determined. i.e. the transducer is employed to determine convection.

2. (B) Force-based methods

   The additional flow of paramagnetic molecules may also be used to push elements or create an additional pressure, both of which may be determined using dedicated apparatus (Pauling *et al* 1946). One possibility is to use a dumbbell construction dangling from a wire. Applying an inhomogeneous magnetic field to the dumbbells will lead to a paramagnetic gas flow to its position thus exerting a force, which in turn will lead to twisting the wire. By determining the twisting angle, the amount of paramagnetic gas may be determined. Likewise, that creation of an additional pressure by means of additional paramagnetic gas flow is feasible.

3.2.1 Selectivity evaluation and miniaturization

Similar to the thermal conductivity sensor, the selectivity is governed by a single parameter characterizing the gas species, in this case the magnetic susceptibility. However, the list of paramagnetic gases is rather short and comprises oxygen (O_2) and nitrogen oxides (NO and NO_2). Of those three, O_2 has by far the largest susceptibility and in most scenarios the oxygen concentration is several orders of magnitudes higher than that of the other two gases. The diamagnetic properties of the other gases may cause minor cross-sensitivities but in most scenarios may be disregarded. In consequence, the selectivity is in practice excellent but then this approach is only useful for those three gases. In fact, sensitivity of paramagnetic sensors limits the applicability to oxygen sensing in practice. In summary, the selectivity is comparable to that of thermal conductivity sensors and the approach may employed in scenarios that require oxygen detection. The use of microsystems technology to construct paramagnetic sensing devices (Schmid *et al* 2006, Vonderschmidt and Müller 2013) is an active research and development field.

3.2.2 Application scenarios

The determination of oxygen is relevant to a number of industrial and medical applications, usually where processes and their safety require a reliable control of relevant parameters including the oxygen concentration. Oftentimes, it is not required to determine trace levels of oxygen but rather higher concentration ranges in the % range.

3.3 Pellistors

Pellistor is a composite word of **pell**et and res**istor** (Gentry and Walsh 1984). Historically it may be viewed as an advancement of the Davy lamp (Davy and Davy 1839) and the catalytic sensor invented by Johnson in 1926. The underlying principle in all cases relies on the burning of flammable gases and a detection of the resulting exothermic reaction. While the Davy lamp becomes brighter, the catalytic sensor detects the increase in temperature resulting from the release of heat and the electronic signal is generated via the temperature dependence of the heater's resistivity. Basically, the catalytic sensor directly converts chemical information into an electric signal via detecting the exothermic oxidation of molecules. As an examplariy reaction the combustion of methane (CH_4) requires oxygen (O_2) and leads to the release of energy via an increase in heat:

$$CH_4 + 2O_2 \rightarrow CO_2 + 2H_2O + Q_{heat}. \tag{3.8}$$

The increase in temperature is proportional to the number of flammable molecules oxidized. It is clear that this reaction requires the availability of oxygen and conditions that allow for the reaction to happen.

Modern day pellistors rely on this basic principle but the construction features a catalytically activated material in order to lower the necessary operation temperature as well as increase the reaction yield. To this end, a platinum wire is surrounded by an insulating metal oxide bead that in turn is decorated with catalytic material. The bead is heated by applying a current to the Pt-wire with the aim to provide the energy to launch the exothermic reaction that oxidizes suitable molecules.

Once this happens, additional heat is deposited into the Pt-wire and its temperature increases. The improvement of performance and reliability follows the strategy of a thermal conductivity detection. A reference channel may be implemented by using a second device at the same temperature but without the catalytic material, i.e. a deactivated pellisor. By comparing the resistivity values of both devices via a Wheatstone bridge only the change in temperature due to additional oxidation is recorded. Since the temperature of the Pt-wire depends on thermal conductivity of the surrounding air (cf thermal conductivity sensor) most setups employing pellistors make use of the reference channel operated at equal temperature. The omission of a catalyst in the metal oxide bead means that it is deactivated for most exothermic reactions at the operation temperature. The setup of a pellistor device is depicted in figure 3.3.

An alternative method to operate pellistors is often aimed at keeping the heating wire's temperature constant and determining the heating power necessary to maintain that temperature. This way, the system is less dependent on ambient temperature changes and the reaction rate is solely dependent on the number of reactive molecules and the availability of oxygen for combustion.

This is an important fact to keep in mind with pellistors: The oxidation of flammable molecules requires oxygen and in atmospheres with high concentrations of flammable or explosive molecules, the lack of oxygen to drive the reaction of equation (3.8) has to be considered. This is usually not an issue when detecting small amounts well below the lower explosive limit (LEL) of the respective target molecules. However, if a pellistor is to be used to detect high concentrations of e.g. methane, then the sensor signal will not be proportional to the number of molecules in air and in fact will decrease for concentration exceeding the upper explosive level (UEL). This is due to a lack of oxidizing gas and consequently, a lower level of converted molecules. In general, the sensitivity of pellistors is closely related to the flammability limits of gas mixtures. The design of micropellistors is an ongoing field and miniaturization is feasible.

3.3.1 Selectivity evaluation and miniaturization

Material science plays a crucial role in improving sensing capabilities of this sensor type as well as reducing size and heating power consumption. Since the pellistor is in fact detecting the rate of oxidizing reactions, its signal is dependent on the parameters of the underlying chemistry. The latter is dependent on temperature, catalytic effects, as well as the amount of heat released per oxidized molecule. This makes the sensitivity function of a pellistor dependent on both, chemistry and temperature, and consequently the selectivity is rather poor since in principle all molecules reacting in an exothermic way are detected. Using the operational temperature of the bead, the selectivity may be tuned slightly, since chemical reactions tend to feature a preferred temperature range. However, the basic working principle already hints at a multitude of cross-sensitivities towards explosive/flammable gases. The points that need consideration include those of thermal conductivity sensors as well as the chemical reactions promoted by the catalysts. Both temperature and catalyst's influence on the selectivity and may be employed to tune the sensor sensitivity towards specific molecules.

In practice, users of pellistor technology should be aware of the type of gases they expect to be present. As an example, the operator of a biogas plant will be interested in monitoring the leakage of methane. To this end, the operating temperature of a pellistor may be tuned to feature the highest sensitivity for this and since other flammable gases are highly unlikely to occur in relevant quantities, the concentration reading of the pellistor device will be fairly reliable.

3.3.2 Application scenarios

Pellistors are usually employed in safety-critical scenarios in industrial environments to monitor the lower explosive limit of gases that may be released from leaks. Oftentimes, the number of gas species that may be released is limited, such that it is possible to tune the operational parameters of the pellistor to optimize sensitivity. One downside of pellistors is the possible poisoning by different classes of molecules. For example, the presence of silicon compounds, such as silicon dioxide (SiO_2), will act a sealant of the porous metal oxide bead resulting in a rapid decrease in sensitivity. This type of drift effects impact on the long-term stability and manufacturers aim to limit the effect by employing filters or other techniques.

3.4 Ionization detectors

The detection of ionized particles via the current they generate may be used as a highly sensitive albeit little selective method for gas detection. Two main types of ionization detectors are readily available and the most important distinction is the method of ionizing molecules. The basic working principle is the same, however, and consists of three steps:

1. Ionize gas molecules.
2. Accelerate the resulting ions in an electric field.
3. Determine the resulting current.

The setups are consequently comprised of an ionization volume, a capacitive setup providing an electric field accelerating ions and determining the resulting current, and a source able to ionize gas molecules. The most common types are so-called flame ionization detectors (William and Dewar 1958) and photoionization detectors (Lovelock 1961). The basic setups for both types are depicted in figure 3.4.

In both cases, a mechanism to create ions is used and the ultimately resulting current is determined. The energy to separate at least one electron from the gas molecules is provided by an ultra-violet photon or chemiionization in a flame.

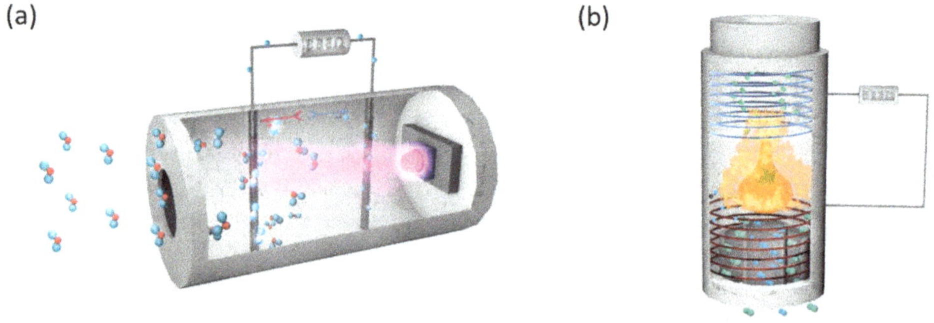

Figure 3.4. Schematic drawing of the working principles of the two main ionization detector technologies. (a) Using an ultra-violet light source one bound electron of a molecule may be removed thus creating an ion and a free electron. Via the use of a static electric field, both particles are accelerated in opposing directions and generate a current. (b) Using a high temperature flame the thermal energy of the gas sample is increased and for a small percentage of the gas molecules, this is sufficient to generate ions.

(A) Photoionization detector (PID)

The means of generating ions in this type of detector (Freedman 1980) is via absorption of a photon with the necessary amount of energy $E_p = \hbar \cdot \omega$ to separate an initially bound electron from the gas molecule. Typically, energy values required for ionization are on the order of 8 eV or above, which translates to photon wavelengths of 155 nm or below. Even though laser sources are in principle available for this spectral range, the use of gas discharge lamps as a light source is much more common due to much lower costs and system complexity. There, high energy photons are generated by using the re-combination processes of a gas plasma and the light is then directed at a gas sample. Noble gas fillings as well as oxygen and hydrogen are commonly used in combination with suitable window materials, such as magnesium fluoride (MgF_2) or calcium fluoride (CaF_2). In a coarse estimation every absorbed photon will generate one molecular ion, which in turn will generate a charge. Hence the measured current I is proportional to the photon absorption rate R_{pa}, the probability of detecting the resulting charge P_d, and the number density of molecules n that may be absorbed:

$$I \sim R_{pa} \cdot P_d \cdot n. \tag{3.9}$$

Two important conclusions about the behaviour of PIDs may be drawn from this, assuming that a constant photon flux, i.e. number of photons per second, is applied:

1. The signal, i.e. the current I, generated by different molecular species depends on the absorption probability of that species. Hence the signal generated by a fixed number density n depends on the gas species and consequently the PID possesses different sensitivities for different molecules.
2. The number of photons along the optical path of the probe volume decreases since every absorbed photon is no longer available for further ionization. Assuming small number density n such that only a small portion of photons is absorbed in total, saturation effects may be neglected and the PID operates in a linear regime and one can assume that ion generation is homogeneous in the whole probe volume. However, once molecular number densities increase to a level that diminishes the photon flux considerably, ion generation will decrease along the optical path. In consequence, the generated current is no longer proportional to the number density for high concentrations.
3. Additionally, a third effect becomes dominant in PIDs for high number densities. Namely, chemical reactions may cause neutralization of the generated ions thus leading to a drop of the signal strength. The reaction routes depend on the availability of oxygen and humidity in the probing volume and are in general manifold. Among them, routes to neutralize the photo-induced molecular ion

are responsible for a diminishing signal. A prerequisite for this is the coincidence of two molecules making the process dependent on n^2 thus dominant for high concentrations.

In summary, PIDs are sensitive in the sense that they allow for determining molecule concentrations in the part-per-billion (ppb) range and they can be used up to fairly high concentrations in the % range. Because of the direct relation between signal and number density in the probing volume they may be used for quantitative analysis, i.e. they do provide a repeatable reading of the concentration. However, the sensitivity of PIDs is a function of the gas molecule species via the probability of photon absorption and consequently PIDs feature poor selectivity. The only means for slightly improving this situation is using different light sources. If the photon energy is insufficient for ionization, the PID will not detect that particular gas species.

3.4.1 Selectivity evaluation and miniaturization

Depending on the molecular potential and the gas temperature the photon energy acts as a threshold value for the generation of ions: if the photon energy is not sufficient to promote the molecule into an ionized state, then the PID setup will not be able to detect this particular molecular species. On the other hand, it will be sensitive to all molecules that may be ionized using this excitation wavelength. The only means to tune the selectivity in this gas detector type is the light source, i.e. the gas used to generate photons in combination with the transmissivity of the window. The most common threshold values for the excitation energy E_p are 9.5, 10.6, and 11.7 eV, corresponding to different groups of gas molecules, namely carbon hydrates, ethanol, and acetylene, respectively. The potential for designing micro-systems is mainly limited by two processes:

1. The absorption of photons scales with the optical path length, i.e. short absorption paths lead to low sensitivity.
2. The number of photons generated in the discharge related to the number of ionized molecules. Both conditions place fundamental restrictions on building ever smaller devices.

3.4.2 Application scenarios

The sensitivity of PID is high and it is possible to detect concentrations in the ppb range. The highest sensitivities are achieved for aromatic hydrocarbons as well sulfidic molecules. It is often used to monitor cancerous such as molecules benzene, toluene, xylene (BTX), which are highly volatile organic compounds used in many processes.

(B)Flame ionization detector (FID)

A slightly less subtle way to produce ions from gaseous molecules is by applying enough thermal energy to ionize molecules (Holm 1997, 1999) and rely on the Boltzmann distribution of energies, which gives the probability W_i of finding a system in a level i with energy E_i at given a temperature T via:

$$W_i \sim e^{-\frac{E_i}{k \cdot T}}. \quad (3.10)$$

This means that by increasing the temperature, the probability of promoting molecules to levels with energies E_i above the ionization threshold increases. In thermal equilibrium at temperatures $T > 1000$ K a sizeable number of molecules is transferred to an ionized state. This may be achieved by flames with a sufficiently high temperature and in fact, the increased conductivity in flames has been observed and employed for at least 70 years.

The fundamental detection mechanism remains the same as in PID detectors, i.e. a capacitor is used to measure the current generated by ions. However, the thermo-chemical processes leading to the generation of ions is more complex and depends on the local temperature and, crucially, on the gas composition that is burned. A common approach is to burn hydrogen gas using air as oxidizer in a blowpipe. The temperature reaches about 2400 K at the zone of highest temperature. The flame chemistry produces different thermal zones and results in various associated ion production processes. The burning of pure hydrogen thus sets the baseline current and mixing the carrier gas hydrogen with the probe gas will change the ion production rate and consequently the current. FIDs are popular for the detection of hydrocarbons, which can be detected in the ppb range using this technique. Depending on the number of carbon atoms of the molecules analyzed, the number of produced ions varies. This means that sensitivity of FIDs depend on the molecular species to a large degree.

3.4.3 Selectivity evaluation and miniaturization

The selectivity of FIDs is poor and comparable to that of pellistors. This is due to the fact that a FID relies on oxidizing the gas molecules to be detected, i.e. they are burned. It also precludes the FID from detection of most inorganic molecules. Among the parameters that influence selectivity is the gas composition. As compared to PIDs the FID approach requires higher instrumental complexity, since the respective gas flows have to be precisely controlled. The potential for miniaturization is limited due the requirements and size of the ionization flame.

3.4.4 Application scenarios

The use cases for FIDs overlap with PIDs and usually they form part of more complex gas sensing systems. The detection of alkanes is a common application for FIDs and the calibration of an FID to provide a quantitative result is fairly straightforward owing to a high degree of linearity between recorded signal and actual number density. Since the chemistry in the flame splits the C–C bonds, the sensitivity per molecules increases with increasing number of carbon atoms in the molecules. Basically, alkane molecules are split to molecules containing a single C, which is then ionized and detected (Cheng *et al* 1998). This means that the sensitivity per molecule for octane is about eight times that of methane, since about eight ions are produced per octane molecule, while a methane molecule is able to produce one ion at most. However, the different alkane molecules may not be distinguished from the sensor signal. Instead more complex apparatus is required to perform a separation prior to a quantification using an FID.

3.5 Metal oxide-based gas sensors

The dependence of electric resistivity of semiconducting layers on the adsorption of molecules on the surface has been known for at least 70 years (Brattain and Bardeen 1953, Heiland 1957). In the following years, metal oxide-based semiconductors have quickly become popular for gas sensing applications, in part because of their high degree of chemical stability. The periodic table of elements provides a vast number of possibilities for producing different metal oxides and one can add to that composite materials and catalyst, that influence the sensing behaviour (Sberveglieri 1992a). However, the popularity of tin dioxide (SnO_2) as well as its paramount historic role makes it an ideal exemplary material to discuss the properties, strengths and weaknesses of metal oxide-based gas sensing without losing generality of the discussion.

Several historic review articles have been published over the years that give a more detailed introduction, and the interested reader is referred to some of them. In the 1950s first Bardeen and Brattain (Brattain and Bardeen 1953) experimenting with germanium and shortly after Heiland (Heiland 1957) using zinc oxide (ZnO) noticed the dependence of the electronic state on the gas composition. Seiyama and Shaver further investigated associated phenomena (Seiyama *et al* 1962, Shaver 1967) and as early as 1970, Taguchi patented a device able to detect gases using SnO_2 as transducer (Taguchi 1971). Even today, metal oxide gas sensors are often referred to as Taguchi-type sensors. The elegance of using the conductivity of metal oxides as transducers is that they directly convert chemical information into an electronic signal, i.e. they constitute a gas sensor without further ado.

The discussion of the processes is intended to give a basic understanding. Readers wishing to dive into the complex processes influencing the behaviour of metal oxide surfaces (Barsan *et al* 2010, Schierbaum *et al* 1991) are referred to dedicated review papers on the topic (Barsan and Weimar 2003). Several books and review papers on the topic are available and a few examples are presented in table 3.4.

The basic signal generation relies on the adsorption of gas molecules on the surface. This is not to be confused with absorption since it is an effect limited to the surface only. Adsorption describes the interaction of gas molecules with only the surface and one can roughly distinguish between so-called physisorption and chemisorption (Masel 1996). Both can be distinguished by their effect on the electronic state of the solid material, the typical bond strength and the average distance between gas molecules and surface. Figure 3.5 depicts a schematic overview of the adsorption types.

Physisorption mostly relies on so-called van der Waals-type interactions between the respective electron distributions in molecule and solid. Consequently, the associated bond is weak ($\sim$10 meV) and does not exhibit a major influence on the resistivity of the metal oxide. The mean distance r from the surface is considerably larger as compared

Table 3.4. A list of possible literature dealing with an in-depth description of the complex process leading to sensor signals using metal oxides as functional materials.

Authors	Title	References
Kohl and Wagner	'Gas Sensing Fundamentals'	Kohl and Wagner (2014)
Bârsan and Weimar	'Understanding the fundamental principles of metal oxide-based gas sensors; the example of CO sensing with SnO_2 sensors in the presence of humidity'	Barsan and Weimar (2003)
Sberveglieri	'Gas sensors—principles, operation, and development'	Sberveglieri (1992b)
Boehme, Weimar and Bârsan	'Unraveling the Surface Chemistry of CO Sensing with In_2O_3 Based Gas Sensors'	Boehme *et al* (2021)
Bârsan, Koziej and Weimar	'Metal oxide-based gas sensor research: how to?'	Barsan *et al* (2007)
Staerz Weimar and Bârsan	'Current state of knowledge on the metal oxide-based gas sensing mechanism'	Staerz *et al* (2022)
Comini	'Solid State Gas Sensing'	Comini (2009)
Bârsan and Schierbaum	'Gas sensors based on conducting metal oxides'	Bârsan and Schierbaum (2019)

to chemisorption. For these reasons, it is not crucially relevant for the working principle of the metal oxide gas sensors and will not be discussed further.

Chemisorption on the other hand may be understood as a chemical bond between the surface and the gas molecule and in fact the typically involved bonding energies are on the order of covalent bonds in a molecule ($\sim$1 eV). The type of bond depends on the respective electronic properties of gas molecules and the surface. In any case, the electronic state of both surface and molecule are fundamentally altered. The basic process of chemisorption is shown in figure 3.6 depicting oxygen as an important reactive gas. The surface of the metal oxide layer features active sites, depicted here as localized centres, which may bond single molecules.

To understand such a system's behaviour a simple model may be developed assuming that:

- each possible adsorption site on the surface may only be occupied once,
- the occupational state of adsorption sites does not alter the behaviour of other sites, i.e. there is no interaction between sites,
- the absorption centres are distributed homogeneously.

With this in mind, the coverage θ of a surface is defined as the ratio of the number of occuped adsorption sites N_o and the number of total adsorption sites N_t:

$$\theta = \frac{N_o}{N_t}, \tag{3.11}$$

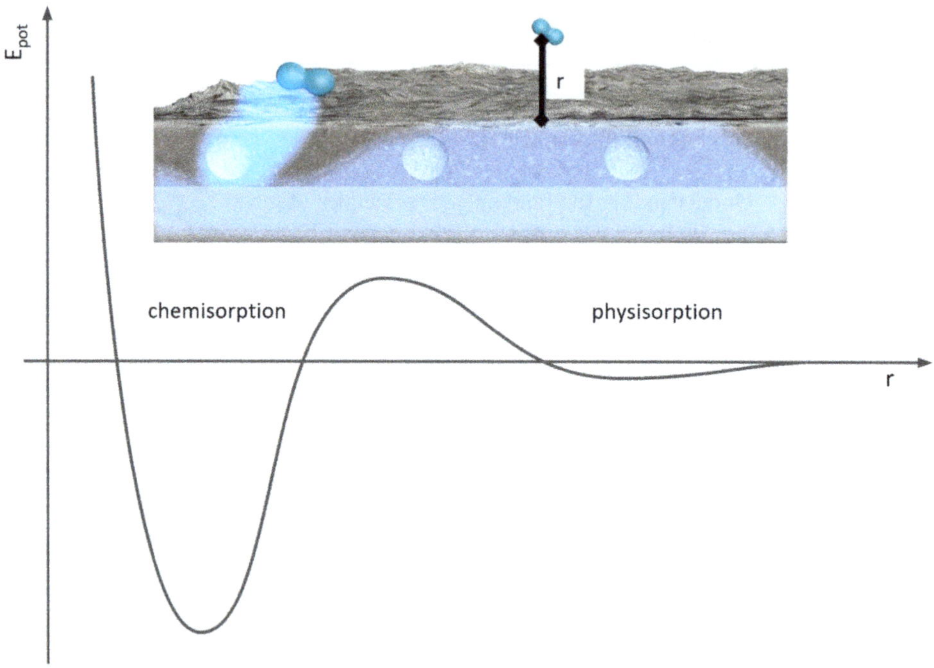

Figure 3.5. The minima in potential energy E_{pot} occur at characteristic distances r between surface and molecule. Physisorption results in a weak bond in a shallow potential minimum between the surface and the gas molecule and does not need to overcome any potential barrier. Chemisorption typically exhibits bonding strengths, that are roughly one order of magnitude larger but potential barriers might have to be overcome. The electronic state of the metal oxide is altered and a finite number of possible surface adsorption sites facilitate the interaction.

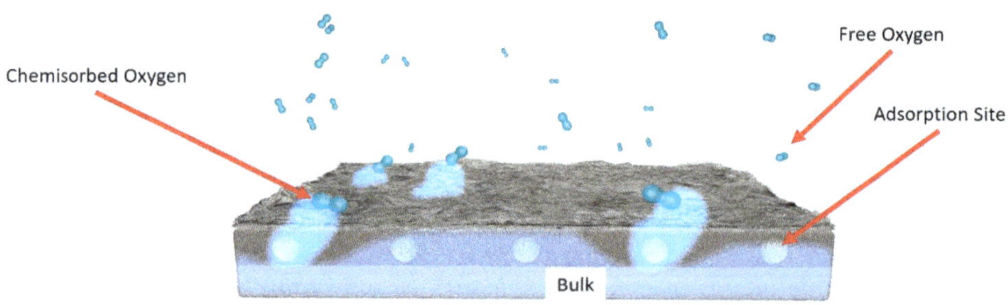

Figure 3.6. The metal oxide surface features a number of possible adsorption sites that may be occupied by a free oxygen molecule from the surrounding air. The resulting bond between surface and molecule pins a charge carrier and thus alters the number of free charge carriers in the functional material, which in turn impacts on its electrical resistivity.

and can take on values between 0 and 1. Any inbound molecule on the surface will chemisorb with a certain probability and a temperature-dependent adsorption rate constant k_{ad}, that depends on the chemical interaction and the temperature described that probability. This leads to a rate of chemisorption R_{ad} that is dependent on the number density of a gas species n_{GS}, the number of free adsorption sites $N_f = N_t - N_o$, and the adsorption rate constant:

$$R_{ad} = n_{GS} \cdot N_f \cdot k_{ad} = n_{GS} \cdot (N_t - N_o) \cdot k_{ad}. \quad (3.12)$$

Likewise, any chemisorbed molecule may undergo a transition to a free molecule. At a fixed temperature the rate depends on the number of occupied sites and the temperature-dependent desorption rate constant k_{des}:

$$R_{des} = N_o \cdot k_{des} \quad (3.13)$$

At a fixed temperature this will lead to an equilibrium, where both rates are equal:

$$R_{des} = R_{ad}$$
$$\Longleftrightarrow$$
$$\theta = \frac{n_{GS}}{\frac{k_{des}}{k_{ad}} + n_{GS}} = \frac{n_{GS} \cdot \frac{k_{ad}}{k_{des}}}{1 + n_{GS} \cdot \frac{k_{ad}}{k_{des}}}. \quad (3.14)$$

Consequently, the temperature-depenent ratio of the adsorption and desorption rate $k = \frac{k_{ad}}{k_{des}}$ constants play a crucial role in the behaviour of metal oxide-based gas sensors. These simplified considerations result in the so-called Langmuir isotherm, which relates the surface coverage to the number density of chemisorbing species for a fixed temperature (Basmadjian 1996). The behaviour of Langmuir isotherms is schematicall depicted in figure 3.7.

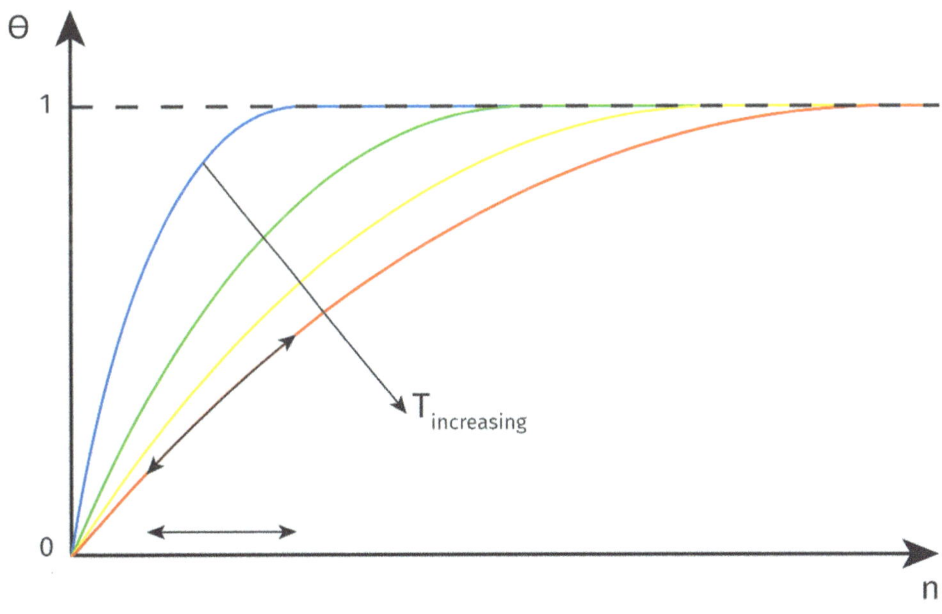

Figure 3.7. The surface coverage as described by the Langmuir isotherm is a function depending on the number density. For a fixed temperature, the surface coverage increases with increasing number density and vice versa. In general, saturation occurs quicker for lower temperatures, since the relative probability of desorption increases with increasing temperature.

Crucially, the surface coverage changes as a function of the number density without hysteresis. In metal oxide-based gas sensing, the coverage is intertwined with the electronic state at the surface. The most important and typically most abundant gas species is oxygen. For understanding the sensing principle an exemplary gas–surface system suffices (Morrison 1987):

In its stochiometric state, tin dioxide (SnO_2) is close to being an isolator with a band gap of 3.6 eV (Batzill and Diebold 2005). In practice, however, it usually exhibits oxygen vacancies that convert the material into an n-type semiconductor since the missing oxygen atoms introduce low lying electron donor states that are depleted by thermal excitation already at room temperature (Kílíç and Zunger 2002). This means that sufficient electrons are available to form a bond when the potential barrier for that bond may be overcome. If an oxygen molecule approaches the surface and the required activation energy is available, then it may form a bond with the surface. The net reaction may then be written as:

$$O_{2,\mathrm{gas}} + e^- \rightarrow O^-_{2,\mathrm{ads}}, \tag{3.15}$$

where $O_{2,\mathrm{gas}}$ denotes free oxygen and $O_{2,\mathrm{ads}}$ chemisorbed oxygen on the SnO_{2-x} surface. This bond requires an electron from the SnO_{2-x} conduction band, which in turn leads to a pinning of that electron and a depletion of free charge carriers in the material. Since the conductivity of n-type semiconducting SnO_{2-x} crucially hinges on the number density and mobility of electrons, the chemisorption of oxygen has a dramatic effect on the resistivity.

Additionally, the influence of chemisorption itself depends on many parameters, including gas species, temperature, and metal oxide material. In a thermodynamic equilibrium, i.e. when operating the metal oxide layer at a constant temperature, adsorption and desorption rates will be equal, and consequently the charge carrier depletion and the electrical conductivity are constant in time but dependent on the oxygen concentration. The latter makes this system a gas sensor. It gives rise to a complex interplay of various parameters making a metal oxide-based gas sensor both an interesting research topic and a difficult to control technology. Under ambient environmental conditions the most prominent gas species adsorbed to the surface are oxygen and water molecules. The latter is usually bound to the surface as hydroxyl group (OH), leading to an increase in electrical conductivity of SnO_{2-x}. The theoretical description of these systems quickly evolves into a highly complex issue, since parameters such as functional layer thickness, surface structure, oxygen vacancy concentration, and crystal structure influence the behaviour and have to be considered. Several books and review papers have dealt with some or all of these issues and the interested reader is referred to some of them in table 3.1. For understanding the strength and weaknesses of a metal oxide-based gas sensor the influence of oxygen, humidity, and temperature provides a sufficient basis.

Oxygen is the most abundant reactive gas in the atmosphere, and its adsorption sets the baseline for metal oxide-based gas sensing. On top of this, humidity has a large effect on the sensor reading. Both molecular species also influence the chemistry at the surface of the MOX layers. Many reactions of gases are in fact indirect via chemisorbed oxygen, such that the change in resisitivity is due to a change in the mean oxygen coverage of the MOX surface. A prominent and often cited example is the detection of carbon monoxide (CO). The CO molecules ultimately react with chemisorbed oxygen and it oxidizes to carbon dioxide (Kohl 1989):

$$2CO_{gas} + O^{-}_{2,ads} \rightarrow 2\ CO_{2,gas} + 2\ e^{-}. \tag{3.16}$$

This alters the steady-state surface coverage of oxygen and in consequence the resistivity is lowered, giving rise to CO sensitivity of MOX-based sensors. Very much in the same way most oxidizing and reducing gases may interact with SnO_{2-x}, but often with much more complex surface reactions. Additionally, the direct adsorption of trace gas species is of course possible thus adding to a complex interplay. Figure 3.8 shows the electrical conductivity of a $SnO_{2\text{-}x}$ layer at different temperatures, an oxygen concentration of 20%, and dry and humid air at 1 bar pressure. It highlights the influence of chemisorption on the electrical conductivity, the influence of temperature on the surface chemistry, as well as the effect of chemisorbed hydroxyl groups.

The measurement highlights the influence on chemisorption and the interplay between temperature and chemical activity. In any case the MOX sensor only knows two directions from its steady-state resistivity reading upon exposure to reactive trace gases, i.e. up or down, at least under nowadays common operation protocols. And this fact finally gives rise to cross-sensitivities. This has been known from the

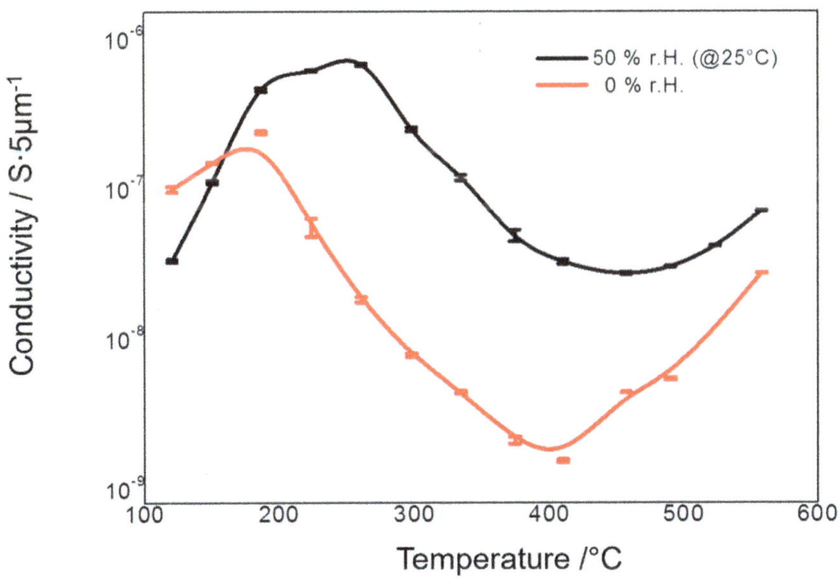

Figure 3.8. The temperature is a decisive factor determining the type and likelihood of chemisorption of oxygen on the surface. The resulting conductivity is shown for dry synthetic air and humid air with a relative humidity of 50% at 25 °C.

very beginning of MOX sensor development and, consequently, much effort has been invested to increase the selectivity of this sensor type. The strategies include the use of different metal oxide materials, thermal modulation, and the use of catalysts to promote selected surface reactions. Additionally, the porosity and thickness of the MOX layer as well as the MOX particles size has a decisive influence on the gas sensing behaviour. A detailed discussion of all influencing factors may be found in dedicated books and review papers. The most important influencing factors are quickly revised in the following.

3.5.1 Metal oxide materials

It is very much possible that all metal oxides are gas sensitive, since chemisorption is a universal process between gaseous molecules and surfaces. Metal oxides exhibit finite electrical conductivity and consequently their resistivity will change. However, the actually deployed materials are not endless and the choice is narrowed down by practical considerations including long-term stability, cost, sensitivity and selectivity. Because of its stability SnO_{2-x} is the most prominent material but since the choice of materials has a considerable impact on selectivity, some of the more important ones include both n-type and p-type materials as well as their combination. Commonly used materials include tungsten oxide (WO_3), titanium oxide (TiO_2), chromium oxide (Cr_2O_3), chromium-substituted titanium oxide (CTO), copper oxide (CuO). The reaction towards different classes of gas molecules differs depending on the material but also as a function of the layer's morphology and porosity. Via combining p- and n-type materials the resulting space charge region can bolster sensitivity.

3.5.2 Catalysts

Similar to their use in pellistors, catalysts are often employed in order to lower operational temperatures as well as enhance selectivity to certain gas species. Usually, noble metals exhibit favourable characteristics, especially when it comes to long-term stability. Among habitual materials are palladium (Pd), platinum (Pt), and gold (Ag).

3.5.3 Thermal modulation

Another aspect of both selectivity and sensitivity is the dependence of the various surface reactions on temperature. Especially when using micromachined sensor systems fast thermal modulation schemes become possible. This in turn may be used to devise alternative operational protocols of tune the temperature in order to increase selectivity (Schütze *et al* 2017).

3.5.4 Microsystems

MOX-based gas sensors are well-suited to building low-cost, miniaturized gas sensing systems using microsystems technology. The fundamental setup requires only an electrode structure and a heating structure to enable layer read-out and temperature control, respectively. The choice of substrate material has no decisive effect and both oxidized silicon and ceramic wafers are currently employed in commercial devices. The metallization is often made of noble metals because of their high chemical and thermal stability. Deposition techniques of the MOX layers range from thick film deposition like screen printing to thin film techniques like sputtering. A review on deposition techniques may be found here (Sahner and Tuller 2010). To lower the power consumption for attaining a certain temperature, the use of chemical etching of Si-wafer to create so-called hotplate devices is often used. A typical setup is shown in figure 3.9 and several companies are producing this type of sensor chip.

It allows for faster thermal modulation while at the same time the power consumption as compared to bulk sensor is lower by about one order of magnitude. Nonetheless, bulk sensors continue to circulate in the market.

In high variability of baseline resistivity and sensitivity, the use of MOX sensors in tasks where quantitative measurements are required is very limited. However, using advanced signal analysis it is possible to use MOX sensors to assess air samples. This is done under the name electronic nose and would require a book of its own.

3.5.5 Selectivity evaluation

The discussion of selectivity in MOX sensors is complex, since it very much depends on the measurement task. For example, hydrogen sulphide may be detected with a high degree in selectivity when using CuO-based materials. However, in general the selectivity of MOX materials is limited and the sensor's performance is characterized by a large number of possible cross-sensitivities, not least due to changing humidity values. On the other hand, the combination of microsystem's technology and a large variety of available and cost-effective materials allows for building sensor arrays, which in turn allows for improving the selectivity.

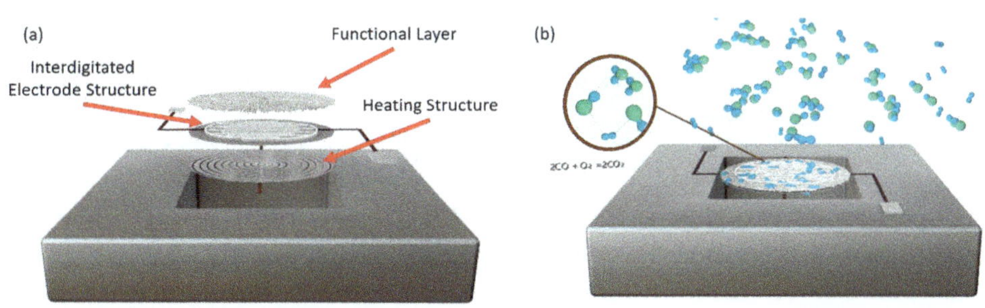

Figure 3.9. (a) Schematic setup of a metal oxide-based gas sensor featuring a buried heating structure and a so-called inter-digitated electrode structure. The functional layer is deposited on top the electrode structure to determine its resistivity. (b) The chemisorbed oxygen can react with trace gases and in consequence the steady-state surface coverage is altered, which leads to a change is resistivity.

3.5.6 Application scenarios

Because of the lack of reproducibility and an additional drift both of the baseline resistivity and the sensitivity over time, the deployment of MOX sensors is currently often limited to scenarios where a quantitative determination is not required. A notable exception is monitoring of hydrogen sulphide in natural gas and biogas applications. There, hydride materials of CuO/SnO_2 are used to detect the concentration in the ppb to ppm range. Apart from that, MOX sensors are discussed for air quality monitoring applications, making use of sensor arrays to assess the concentration of volatile organic compounds. Also, monitoring of the LEL in households using gas to cook is a possible usage scenario. Finally, the so-called electronic noses often use MOX gas sensors to evaluate odour or other scenarios.

3.6 Electrochemical cell

The working principle of a so-called electrochemical cell may be pictured as an inverse battery and it features the same basic components, namely an ion conductor and two electrodes (Cao *et al* 1992). Each electrode forms a half-cell in combination with the electrolyte and particular chemical reactions occur. While a complete description of the signal generation process is complex, the net reactions leading to an electrical current give sufficient insight to evaluate this technique.

The whole idea of an electrochemical cell relies on a chemically induced potential difference between two electrodes. To this end, one electrode is to be constructed to be gas permeable, while containing the ion conductor, which often is a liquid. Additionally, this so-called working electrode is usually made of a catalytic material in order to promote the chemical reactions necessary for gas detection. In practice this is usually a porous layer of platinum or gold. Typically, the gas to be detected is oxidized at the working electrode when electrolyte and gas intersect and charges are separated in the process, as indicated in an exemplary working electrode reaction:

$$CO + H_2O \rightarrow CO_2 + 2\,H^+ + 2\,e^-. \qquad (3.17)$$

The ions are transported via the electrolyte, while the electrons are transported in a conductor to the so-called counter electrode. The latter constitutes the transducing mechanism and from equation (3.10) it becomes apparent that the current detected is a direct measure of the number of molecules that have been oxidized at the working electrode. The counter electrode closes the loop by re-combining the generated charges using the net reaction:

$$2\,H^+ + O_2 + 2\,e^- \rightarrow H_2O. \qquad (3.18)$$

The combined net reaction at both electrodes is hence made up of oxidation of carbon monoxide to carbon dioxide. This type of sensing directly converts chemical information into an electronic signal, namely a current. The number of charges detected per second is directly proportional to the number of oxidized gas molecules per second. This rate, however, is influenced by a number of processes such as the diffusion through the porous membrane and the temperature of the system figure 3.10.

Figure 3.10. (a) Schematic setup of an electrochemical gas sensor with a working electrode that allows diffusion of gas into the cell. The resulting chemical potential provokes a flow of ions in the electrolyte and the electrons to balance the charge are counted to provide a reading of the gas concentration.

3.6.1 Selectivity evaluation and miniaturization

The selectivity is governed by the specific electrochemical potential of the respective cell as well as the diffusion properties of the porous membrane. In particular, electrochemical cells employing a water-based electrolyte are prone to cross-sensitivities, e.g. to changes in humidity. Manufacturers aim to reduce this influence by employing filters but the diffusion times through the membrane then limits the reaction speed. Other than that, the selectivity of electrochemical cells is good and interfering gases are often time-suppressed by a least a factor of 100. Currently, liquid-based electrolytes limit the potential for miniaturization but examples of solid-state ion-conducting materials also highlight the potential for using a micro-system's engineering to construct miniature electrochemical gas sensing devices.

3.6.2 Application scenarios

The most used electrochemical sensor is probably the so-called lambda sensor, an oxygen sensor based on a solid-state ion conductor used to determine the oxygen content of the exhaust of combustion processes. It effectively determines the oxygen concentration gradient between ambient air and the exhaust gas.

References

Barsan N, Koziej D and Weimar U 2007 Metal oxide-based gas sensor research: how to? *Sens. Actuators* B **121** 18–35

Barsan N, Simion C, Heine T, Pokhrel S and Weimar U 2010 Modeling of sensing and transduction for p-type semiconducting metal oxide based gas sensors *J. Electroceram.* **25** 11–9

Bârsan N and Weimar U 2003 Understanding the fundamental principles of metal oxide based gas sensors; the example of CO sensing with SnO_2 sensors in the presence of humidity *J. Phys.: Condens. Matter* **15** R813–39

Bârsan N and Schierbaum K 2019 *Gas Sensors Based on Conducting Metal Oxides. Basic Understanding, Technology and Applications* (Metal Oxides Series) (Amsterdam: Elsevier) https://sciencedirect.com/science/book/9780128112243

Basmadjian D 1996 *The Little Adsorption Book. A Practical Guide for Engineers and Scientists* 1st edn (Boca Raton, FL: CRC Press) https://search.ebscohost.com/login.aspx?direct=true&scope=site&db=nlebk&db=nlabk&AN=1707874

Batzill M and Diebold U 2005 The surface and materials science of tin oxide *Prog. Surf. Sci.* **79** 47–154

Böckh P and Wetzel T 2012 *Heat Transfer: Basics and Practice* (Berlin: Springer)

Boehme I, Weimar U and Barsan N 2021 Unraveling the surface chemistry of CO sensing with In_2O_3 based gas sensors *Sens. Actuators* B **326** 129004

Boltzmann L 1884 Ableitung des Stefan'schen Gesetzes, betreffend die Abhängigkeit der Wärmestrahlung von der Temperatur aus der electromagnetischen Lichttheorie *Ann. Phys.* **258** 291–4

Brattain W H and Bardeen J 1953 Surface properties of germanium *Bell Syst. Tech. J.* **32** 1–41

Cao Z, Buttner W J and Stetter J R 1992 The properties and applications of amperometric gas sensors *Electroanalysis* **4** 253–66

Carlin R L and van Duyneveldt A J 1977 Paramagnetism: the curie law *Magnetic Properties of Transition Metal Compounds, Bd. 2* (Inorganic Chemistry Concepts, 2) ed L Richard, A J Carlinund and H Duyneveldt (Berlin: Springer) pp 1–22

Chapman S and Cowling T G 1998 The mathematical theory of non-uniform gases. An account of the kinetic theory of viscosity *Thermal Conduction and Diffusion in Gases* (Cambridge Mathematical Library) 3rd edn (Cambridge: Cambridge University Press) (Transferred to digital printing)

Cheng W K, Summers T and Collings N 1998 The fast-response flame ionization detector *Prog. Energy Combust. Sci.* **24** 89–124

Comini (Hg.) E 2009 *Solid State Gas Sensing* (Boston, MA: Springer)

Davy H and Davy J 1839 *The Collected Works of Sir Humphry Davy* (London: Smith, Elder and Co.)

Daynes H A 1933 *Gas Analysis: By Measurement of Thermal Conductivity* (University Press)

Dyer C A 1947 A paramagnetic oxygen analyzer *Rev. Sci. Instrum.* **18** 696–702

Freedman A N 1980 The photoionization detector *J. Chromatogr.* A **190** 263–73

Gardner E L W, Gardner J W and Udrea F 2023 Micromachined thermal gas sensors—a review *Sensors* **23** 681

Gentry S J and Walsh P T 1984 Poison-resistant catalytic flammable-gas sensing elements *Sens. Actuators* **5** 239–51

Heiland G 1957 Zum Einflu von Wasserstoff auf die elektrische Leitfhigkeit an der Oberflche von Zinkoxydkristallen *Z. Phys.* **148** 15–27

Hellmann R, Bich E, Vogel E, Dickinson A S and Vesovic V 2009 Calculation of the transport and relaxation properties of methane. II. Thermal conductivity, thermomagnetic effects, volume viscosity, and nuclear-spin relaxation *J. Chem. Phys.* **130** 124309

Hiller J M and Baldwin N M 2005 Thermal conductivity detectors *Environmental Instrumentation and Analysis Handbook* ed D Randy, H Downund Jay and H Lehr (Hoboken, NJ: Wiley-Interscience) pp 417–31

Holm T 1997 Mechanism of the flame ionization detector II. Isotope effects and heteroatom effects *J. Chromatogr.* A **782** 81–6

Holm T 1999 Aspects of the mechanism of the flame ionization detector *J. Chromatogr.* A **842** 221–7

Jasek K, Pasternak M and Grabka M 2022 Paramagnetic sensors for the determination of oxygen concentration in gas mixtures *ACS Sens.* **7** 3228–42

Kílíç C and Zunger A 2002 Origins of coexistence of conductivity and transparency in SnO_2 *Phys. Rev. Lett.* **88** 95501

Kohl C-D and Wagner T 2014 Gas sensing fundamentals *Unter Mitarbeit von Claus-Dieter Kohl* (Springer Series on Chemical Sensors and Biosensors Series, vol 15) 1st edn (Berlin: Springer) https://ebookcentral.proquest.com/lib/kxp/detail.action?docID=1802738

Kohl D 1989 Surface processes in the detection of reducing gases with SnO_2-based devices *Sens. Actuators* **18** 71–113

Lemmon E W and Jacobsen R T 2004 Viscosity and thermal conductivity equations for nitrogen, oxygen, argon, and air *Int. J. Thermophys.* **25** 21–69

Lovelock J E 1961 Ionization methods for the analysis of gases and vapors *Anal. Chem.* **33** 162–78

Marsh K N, Perkins R A and Ramires M L V 2002 Measurement and correlation of the thermal conductivity of propane from 86 K to 600 K at pressures to 70 MPa *J. Chem. Eng. Data* **47** 932–40

Masel R I 1996 *Principles of Adsorption and Reaction on Solid Surfaces* (Wiley Series in Chemical Engineering) (New York: Wiley) http://loc.gov/catdir/bios/wiley047/95017776.html

Mehl J B, Huber M L and Harvey A H 2010 Ab initio transport coefficients of gaseous hydrogen *Int. J. Thermophys.* **31** 740–55

Mehra J and Rechenberg H 1982 *The Historical Development of Quantum Theory. Its Foundation and the Rise of Its Difficulties; 1900–1925* (New York: Springer)

Morrison S R 1987 Mechanism of semiconductor gas sensor operation *Sens. Actuators* **11** 283–7

Pauling L, Wood R E and Sturdivant J H 1946 An instrument for determining the partial pressure of oxygen in a gas *Science* **103** 338

Planck M 1989 The theory of heat radiation *History of Modern Physics, 1800–1950* (Los Angeles, CA: Tomash; American Institute of Physics) http://loc.gov/catdir/enhancements/fy1306/88034240-d.html

Poole C F 2005 Gas chromatography | detectors *Encyclopedia of Analytical Science* 2nd edn ed A Townshend, C F Poole and P Worsfold (Hg.) (Amsterdam: Elsevier) pp 95–105 https://sciencedirect.com/science/article/pii/B0123693977002223

Rumble (Hg.) J R 2022 *CRC Handbook of Chemistry and Physics. A Ready-Reference Book of Chemical and Physical Data* 103rd edn (Boca Raton, FL: CRC Press)

Saggion A, Faraldo R and Pierno M 2019 *Thermodynamics. Fundamental Principles and Applications* (Cham: Springer) (UNITEXT for Physics)

Sahner K and Tuller H L 2010 Novel deposition techniques for metal oxide: prospects for gas sensing *J. Electroceram.* **24** 177–99

Sberveglieri G 1992a *Gas Sensors. Principles, Operation, and Development* (Dordrecht: Kluwer)

Sberveglieri (Hg.) G 1992b *Gas Sensors: Principles, Operation, and Development: Conf. Papers* (Dordrecht: Kluwer)

Schierbaum K D, Weimar U, Göpel W and Kowalkowski R 1991 Conductance, work function and catalytic activity of SnO_2-based gas sensors *Sens. Actuators* B **3** 205–14

Schmid U, Seidel H, Mueller G and Becker T 2006 Theoretical considerations on the design of a miniaturised paramagnetic oxygen sensor *Sens. Actuators* B **116** 213–20

Schütze A, Baur T, Leidinger M, Reimringer W, Jung R, Conrad T and Sauerwald T 2017 Highly sensitive and selective VOC sensor systems based on semiconductor gas sensors: how to? *Environments* **4** 20

Seiyama T, Kato A, Fujiishi K and Nagatani M 1962 A new detector for gaseous components using semiconductive thin films *Anal. Chem.* **34** 1502–3

Shaver P J 1967 Activated tungsten oxide gas detectors *Appl. Phys. Lett.* **11** 255–7

Staerz A, Weimar U and Barsan N 2022 Current state of knowledge on the metal oxide based gas sensing mechanism *Sens. Actuators* B **358** 131531

Stewart S M and Johnson R B 2017 *Blackbody Radiation. A History of Thermal Radiation Computational Aids and Numerical Methods* (Optical Sciences and Applications of Light) (Boca Raton, FL: CRC Press, Taylor and Francis Group)

Taguchi 1971 Gas-detecting device *Google Patents.*

Thodos G 1965 Thermal conductivity of gases and liquids, N. V. Tsederberg, Massachusetts Institute of Technology Press, Cambridge (1965). 246 pages *AlChE J.* **11** 770

Vesovic V, Wakeham W A, Olchowy G A, Sengers J V, Watson J T R and Millat J 1990 The transport properties of carbon dioxide *J. Phys. Chem. Ref. Data* **19** 763–808

Vonderschmidt S and Müller J 2013 A fluidic bridge based MEMS paramagnetic oxygen sensor *Sens. Actuators* B **188** 22–30

William I G and Dewar R A 1958 *Proceedings of the 2nd Symposium on Gas Chromatography. Discussion Group* ed D H Desty (Butterworth)

Zhao B 2022 Integrity of Newton's cooling law based on thermal convection theory of heat transfer and entropy transfer *Sci. Rep.* **12** 16292

IOP Publishing

Gas Sensor Technologies for Environmental Sensing

Stefan Palzer

Chapter 4

Spectroscopic methods

The interaction of electromagnetic waves and gas molecules is manifold and includes elastic and inelastic scattering processes as well as absorption. While highly sophisticated spectroscopic methods have been developed in the laboratory and allow for controlling matter down to the internal and external quantum states, the discussion here is limited to the most basic interactions and setups as they are used in common applications. Besides the possibility for highly selective determination of individual gas species and even isotope-selective sensing, optical methods do allow for contactless and stand-off sensing. This makes it highly attractive for inhospitable environments. This is possible because the ultimate transducing device is not in direct contact with the analyte.

The word spectroscopy derives from the latin word spectrum, which translates to appearance, form, or image. Isaac Newton was probably the first to use the term in connection with colours appearing upon illuminating a prism with white light (Newton 2012). His first scientific manuscript described his findings and also showed that the visible spectrum may be recombined into white light (Mills 1981). However, it is still unclear why Newton did not report the now so-called Fraunhofer lines. By most accounts they should have been visible to Newton in 1666 but it was not until 1802 that Wollaston (Wollaston 1802) reported some missing colours in an otherwise continuous spectrum. Later, Fraunhofer used a diffraction grating instead of a prism and managed to obain somewhat more precise results and higher resolution (Hearnshaw 1990). The reason for the missing lines in the visible spectrum of the sun is the absorption of light by matter and the governing principle of absorption spectroscopy. The 19th century brought several important advances in terms of understanding this interaction (James 2009). Notably, Herschel and Talbot observed discrete emission lines from flames and shortly after Ångström observed that emission lines of hot gases and absorption lines do occur at the same wavelength (Hearnshaw 1990, Hearnshaw 2014). The reason for these phenomena is the interaction of light with the internal states of molecules, which are discussed in

doi:10.1088/978-0-7503-3159-3ch4

Table 4.1. Examples of texts that provide an in-depth description of the scientific fields this chapter is concerned with.

Topic	Author and title	References
Optics and Photonics	Saleh/Teich: 'Fundamentals of Photonics'	Saleh and Teich (2001)
Optics and Photonics	Hecht: 'Optics'	Hecht (2017)
Molecular structure/ Quantum mechanics	Gray: 'Chemical Bonds: An Introduction to Atomic and Molecular Structure'	Gray (1995)
Molecular structure/ Quantum mechanics	Aruldhas: 'Molecular Structure and Spectroscopy'	Aruldhas (2014)
Spectroscopy	Svanberg: 'Atomic and Molecular Spectroscopy: Basic Aspects and Practical Applications'	Svanberg (2022)
Spectroscopy	Mallick: 'Fundamentals of Molecular Spectroscopy'	Mallick (2023)
Spectroscopy	Demtröder: 'Laser Spectroscopy: Vol. 1 Basic Principles'	Demtröder (2008)

the second chapter. While some techniques discussed in the last chapter use the effects that the structure of internal levels has on the macroscopic behaviour, spectroscopy accesses the internal level directly. Table 4.1 gives a short list of possible texts that provide a detailed introduction to the main underlying topics of this chapter, namely the theory of light and photonics, molecular structures and internal energy levels, and spectroscopy as a tool to investigate matter.

The development of a comprehensive theory of light and its behaviour is closely linked with that of matter. Here, a coarse model of light shall be used to explain the effects used in building gas detectors based on the interaction between light and matter. The model will borrow aspects of quantum mechanics as well as the classical description of light and the main magnitudes characterizing light in a quantum description are introduced below.

4.1 Properties of light

Energy

The energy of light in a classical description is governed by the respective amplitudes of the electric and magnetic fields. In the quantum description of light its energy may not attain arbitrary values but there is an indivisible, smallest unit which is called a photon (Lewis 1926). The energy associated with a photon E_P is determined by its frequency υ or wavelength λ via the relation (Saleh and Teich 2001):

$$E_P = h \cdot \nu = h \cdot \frac{c_0}{\lambda}. \tag{4.1}$$

Again, h is Planck's constant and the relation connects the classical magnitude of wavelength with the quantum mechanical description of light.

Momentum

Even though a photon does not possess mass, it does carry a momentum p, which is associated with the classical magnitude of the wavevector $\vec{k}$, with $|\vec{k}| = \frac{2\pi}{\lambda}$:

$$\vec{p} = \hbar \cdot \vec{k}. \tag{4.2}$$

Spin

A photon also carries a spin (Raman and Bhagavantam 1931, 1932) possessing the unit of an angular momentum. It is connected to the polarization of a light wave in the classical description of light. Namely, right circularly and left circularly polarized light are described by:

$$\vec{S} = \frac{+}{-}\frac{h}{2\pi} = \frac{+}{-}\hbar, \tag{4.3}$$

respectively. The classical linear polarizations of light correspond to superpositions of those two spin states. Figure 4.1 depicts a photon and its three fundamental properties. The relation to the classical properties is summarized in table 4.2.

These properties of a photon cannot be separated from each other, which is important to keep in mind for the remainder of the discussion in this chapter. This means that if a photon interacts with matter, these magnitudes interact all at once or not at all. When looking at the complete system, i.e. matter and light, then all of these magnitudes must be preserved. If a photon is absorbed by matter, then its energy, momentum, and spin must be transferred to the matter it has interacted with.

Readers wishing to read up on the theoretical background of most of the possibilities of interaction between photons and molecules are referred to Saleh

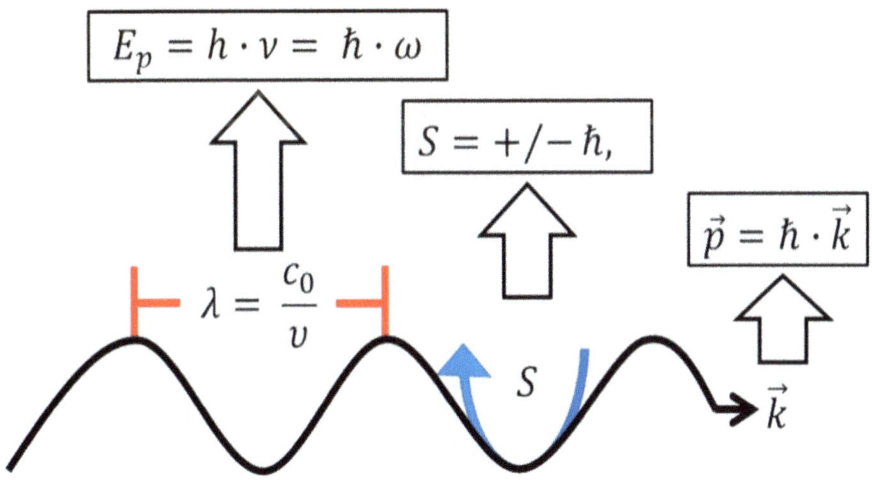

Figure 4.1. A photon travels at the speed of light in a direction given by the wavevector $\vec{k}$ and possesses energy E_P, momentum $\vec{p}$, and spin S, whose value may be either $+\hbar$ or $-\hbar$. The three properties may not be separated from the photon, such that interaction with molecules will always have to be accompanied by a simultaneous transfer of all three at once.

Table 4.2. A comparison between the concepts of classical electromagnetic theory and the corresponding quantum properies of photons.

Electromagnetic wave theory	Quantum optics
Radiant energy	Number of photons
Radiation pressure	Momentum of photons
Polarization	Spin

and Teich (2001), Weiner and Ho (2007), Reddy (2009), and Demtröder (2015). This discussion, however, is limited to absorption aided by so-called dipole-allowed transitions, which summarize all those interactions with a high probability of photon absorption via interacting with electronic dipoles of molecules. The underlying process for this movement may be found in molecular transitions of electronic, vibrational, and rotational nature and ultimately these are responsible for the characteristic absorption features of molecules. Since the energy levels depend on the molecular composition, i.e. which atoms are bound to each other and how, it is also possible to deduce the molecule type by measuring its absorption spectrum. The range of energies associated with the different transitions also determines the range of frequencies of light one has to use to probe that transition.

This places a couple of conditions on the combined system, since overall energy, momentum, and angular moment have to be conserved. Consequently, a molecule absorbing a photon has to assume the photon's momentum, energy, and spin (angular momentum) during the process.

4.2 Properties of matter

Since gas molecules may be considered free particles, no relevant restrictions with respect to the photon momentum need discussion here.

Upon absorption, the photon's energy is transferred to the discrete internal energy levels of matter. In turn, this means that matter may not absorb arbitrary photon energies but only photons with an energy equal to the energy gap between two internal energy levels. A photon may be absorbed by a molecule only in case the energy difference ΔE between two internal energy levels equals the photon's energy E_p, i.e.:

$$\Delta E = E_2 - E_1 = h \cdot v = E_p. \tag{4.4}$$

In case of absorption the photon disappears and all its characteristics are transferred to the molecule.

For the photon being able to interact with matter its spin has to be transferred simultaneously. After all, not all energy levels of a molecule may be easily addressed using a single photon due to restrictions imposed by the angular momenta of the internal molecular levels. This gives rise to the so-called selection rules, which are at the heart of the structure of spectroscopic features employed in gas sensing. This means that the molecule's total angular momentum has to change by $+/-\hbar$.

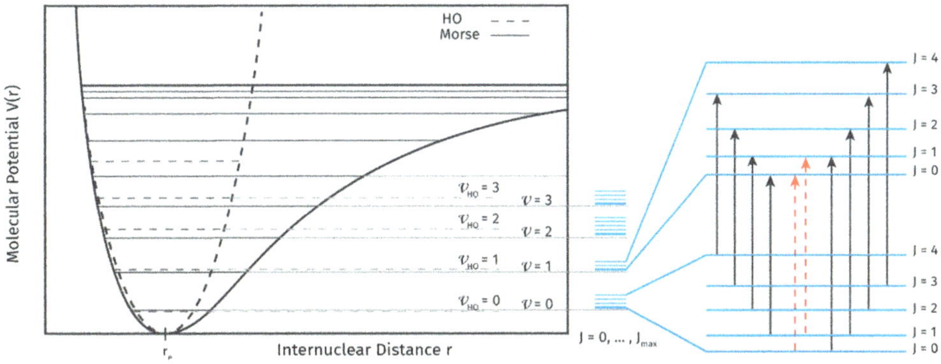

Figure 4.2. Schematic of a vibrational potential and the vibrational energy levels which are a result of the quantization of the possible vibrational energy states. On top of the vibrational levels the rotational levels, whose associated energies are typically much smaller.

To demonstrate the consequence of this, the transition between two vibrational states in a molecule are instructive. Because the rotational energies are much smaller, each of the vibrational states is modulated by many rotational states. Figure 4.2 illustrates the potential of a molecule, the resulting vibrational quantum levels as well the rotational levels. The electronic state of the molecule remains unchanged in this scenario and the typically involved energies here lead to photons with wavelengths in the infrared range to have the right amount of energy to connect two vibrational states. Assuming that the vibration does not contain any angular momentum, absorption of the photon is only possible if a rotational transition takes over the photon's spin. Because the photon only possesses one $\hbar$ spin, the rotational quantum number J may only be changed to $J - 1$ or $J + 1$. No other transitions are possible and as a consequence, only those transitions are observed. The $J - 1$ transitions are called the P-branch, while the $J + 1$ transitions belong to the R-branch. Depending on the type of vibration, one further type of transition is possible, i.e. the Q-branch, where the rotational quantum number of the molecules remains unchanged. A much more precise theoretical description of the emergence of infrared spectra and the selection rules may be found in Quack and Merkt (2011).

In summary, the interaction between light and molecules is governed by their respective quantum states. Whether a photon is likely to be absorbed by a specific molecule species is a function of the internal state of the molecule and the characteristics of the photon. In particular, the molecule has to be able to absorb energy, moment, and spin of the photon at once. The so-called selection rules determine which two internal states of the molecule may be connected via the absorption of a photon:

- The first condition is easy to fulfill since the molecules are freely moving particles, thus they can absorb any momentum the photon might have.
- The second condition is summarized in equation (4.4) and basically defines the photon's frequency.
- The third condition gives rise to the selection rules since the molecule has to absorb the photon's spin.

4.3 Interaction of light and matter

This short introduction allows for using a simple model to describe the interaction between light and matter. Since photon properties may not be separated, it can only interact with two energy levels of a gas molecule. Therefore, one can describe the situation using two energy levels of a molecule that may interact with one photon. The corresponding drawing is presented in figure 4.3.

This model allows for evaluating the processes that might occur when photons and molecules interact. The energy levels in this model represent any two internal levels of the molecule as described in chapter 2, i.e. electronic, vibrational, and rotational states including their combinations. Throughout the discussion of this simple description, only one photon may interact with a molecule at a given time. In effect this means that two photon processes or more complex interactions are disregarded, even though they are possible. However, for understanding the gas sensing techniques discussed in this text they are not required.

After photon absorption, the excited molecule may re-emit a photon some time after without the influence of further photons (Saleh and Teich 2001). This process is called spontaneous emission and the average time until re-emission is mainly a function of the energy difference between excited and ground state. The more energy stored in the molecule, the quicker the spontaneous emission occurs. The spontaneously emitted photon now contains energy, momentum, and spin the molecules lose during this process. However, the direction of re-emission is arbitrary, which in consequence means that only a small part of the photons is re-emitted in the original direction of travel of the original photon.

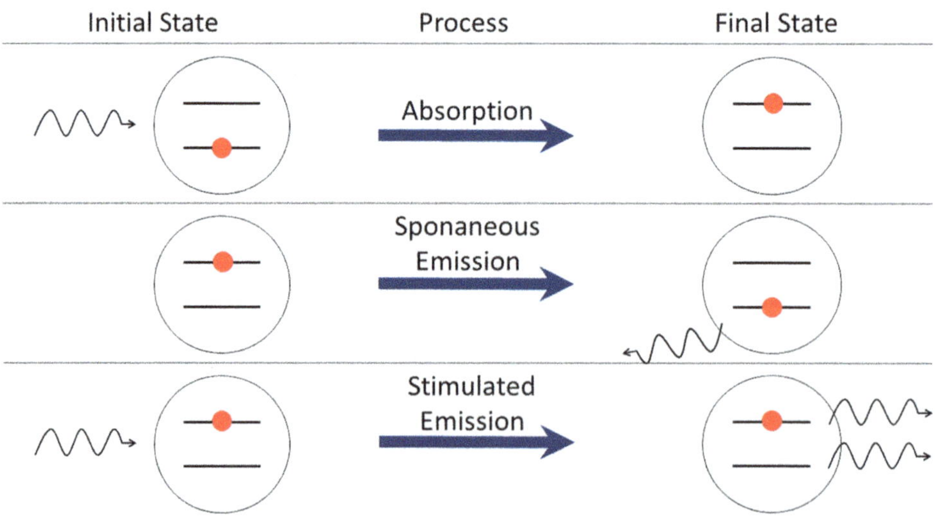

Figure 4.3. A two-level system represents the molecule, where each energy level might be an electronic, vibration, rotational energy state or an arbitrary combination of those. The photon can interact with the two-level system in case it is possible to deposit all three properties during the interaction. The total energy, momentum, and angular moment are preserved at all times.

4.4 Tools of the trade

If a photon satisfying the conditions for absorption coincides with a molecule in the excited state the so-called stimulated emission may happen (Saleh and Teich 2001). In this case, the molecule undergoes a transition from excited state to ground state and a second photon is emitted, which is a precise copy of the photon causing the process. The clone photon is equal to the original photon in all aspects, notably including direction, energy, spin, and momentum.

The various techniques discussed in this chapter all rely on the absorption of photons by molecules and are enabled by the fact that spontaneous emission leads to photons being emitted in an arbitrary direction. The setup of absorption spectroscopy devices is in principle simple and depicted in figure 4.4. The main components are:

(A) a light source,
(B) a light detector to quantify the amount of light after interaction, and
(C) an optical path where gas molecules and light can interact.

However, there are many different possible architectures to build such a sensor and sensor operation, setup complexity and selectivity do depend on the properties of the individual components with respect to different frequencies of the electromagnetic wave utilized.

All absorption spectroscopy setups have the basic working principle in common, that a portion of light sent out by the light source is missing at the detector because it has been absorbed by gas molecules. The different techniques based on absorption spectroscopy are mainly distinguished by the spectral bandwidth of the light source

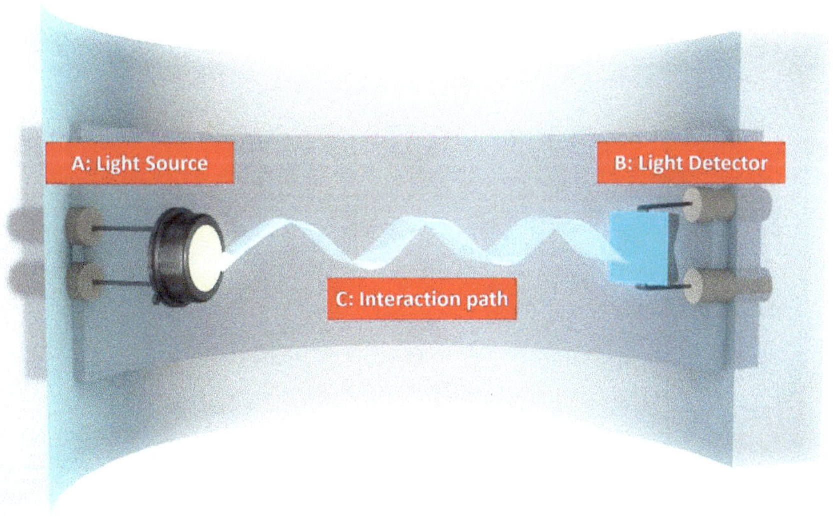

Figure 4.4. The fundamental setup of sensors based on absorption spectroscopy requires a light source (A), an optical path where molecules and light can interact (C), and a light detector (B) to quantify the amount of light after the interaction. The attenuation of the light intensity is the key parameter to determine the number density of gas molecules.

and the way spectral information about molecules is extracted. To gain a feeling for this a short overview of light sources and light detector is helpful.

4.4.1 Light sources

Emission of photons is a fundamental property of all matter at temperatures above 0 K and it has been discussed as heat loss mechanism in chapter 2. The heating up of a surface may consequently also be used as a light source for spectroscopic systems. To highlight this, figure 4.5 shows the spectral distribution of an ideal black body for several temperatures according to Planck's law.

Building a thermal emitter is not complex in principle, since Joule heating of a wire will do the trick already. In fact, miniturized light bulbs are commonly used to build instruments based on absoprtion spectroscopy. The actually emitted spectral distribution depends on the material that is heated up as well as further components of the setup, which might absorb some parts of the spectrum more strongly than others. Besides light bulbs, micromachined, planar emitters are often employed in order to enable an increase of the thermal modulation frequency and decreased power consumption. Additionally, restricting the spectral emission of thermal emitters to desired regions via structuring of the surface is an ongoing research field. However, one has to weigh up between reducing the power consumption and

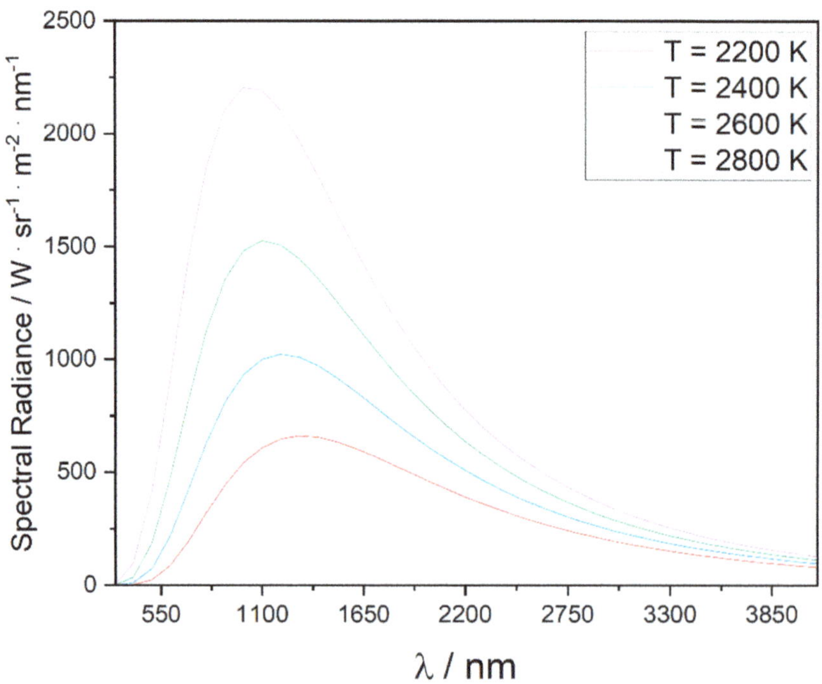

Figure 4.5. The spectral distribution of light emitted by a surface at temperature T in the range from 300 nm to 4.2 μm. Notably, none of the curves cross at any point and the maximum of spectral radiance is dependent on the temperature.

the radiation power emitted, such that the physics of the black body will place contrains on the power consumption achievable.

Lasers

While thermal emitters are broad band light sources basically being able to cover the complete electromagnetic spectrum, a laser can be considered the other extreme in terms of spectral bandwidth. Laser is an acronym for **l**ight **a**mplification by **s**timulated **e**mission of **r**adiation, and its working principle is based on stimulated emission of photons. The ability to clone photons via the process of stimulated emission is the key to building spectrally narrow light sources. To build a laser three components are necessary:

1. A so-called active medium, which is a physical system that allows for the **amplification** of light via stimulated emission. The energy level structure of this system is such that it is possible to create an inversion of the population of the energy levels, which means that it is more likely to find the system in the excited state than in a ground state. In this scenario a photon entering the active medium is more likely to create a clone of itself than being absorbed. This in effect causes the amplification of light.
2. To achieve population inversion some sort of **energy supply** is necessary, which in the context of lasers is often refered to as a pump. Pumping the active medium is the prerequisite for achieving laser radiation.
3. An optical resonator fullfills two tasks: It **recycles photons** that otherwise would have been lost from the active medium and it **selects the frequencies** the laser can emit. Because any photons that are not resonant to the cavity experience high losses, those frequencies will not be dominant.

There is a large number of possible realizations of a laser, including solid-state lasers, gas lasers, dye lasers, and diode lasers. The latter is the most relevant type of laser for gas sensing. It uses a semiconducting material with a bandgap E_g and a p- and n-doped layer as active medium and and a current as pump. The facets of the active medium serve as mirrors.

Light emitting diodes

A p/n junction of a direct semiconductor featuring radiative recombination of electron–hole pairs is used to build light-emitting diodes (LEDs). The semiconducting material's bandgap structure defines the spectral emission profile and creation of electron/hole pairs is achieved via applying a current. An LED is basically a diode laser without the mirrors to sustain significant levels of stimulated emission. The spectral width is many orders of magnitude broader than that of a laser but still much narrower than that of a thermal emitter. It is mainly governed by the bandgap and the occupancy of quantum levels (Saleh and Teich 2001). As opposed to thermal emitters, it is possible to achieve high modulation frequencies in the MHz range as compared to thermal emitters.

Gas discharge lamps

Lastly, gas discharge lamps are also used as light sources in absorption spectrocopy. They have been introduced already earlier during the discussion of PIDs and they serve mainly as light sources for applications where ultra-violet (UV) photons are needed. The frequency of the emitted light is determined by the choice of atomic species and the spectral bandwidth. An overview of basic properties of different classes of light sources is summarized in table 4.3.

4.4.2 Light detectors

Ultimately, the light intensity has to be measured by converting light into an electronic signal. This may be achieved by two fundamentally different approaches, which are distinguished by how the electric signal is created. While photonic light detectors rely on the conversion of a photon into an electron–hole pair, thermal light detectors convert photons to heat (figure 4.6).

Photonic detectors

The absorption of a photon by a semiconducting material may lead to the creation of electron–hole pairs if the photon energy is larger than the bandgap energy. This in

Table 4.3. A coarse overview of the attributes of the most important light sources that are employed in spectroscopic setups. Technical requirements typically determine the choice of the light source.

	Laser	LED	Thermal	Gas discharge
Working principle	Semiconductor	Semiconductor	Joule heating	Gas discharge
Spectral bandwidth	Quasi-monochromatic	Few nm–1 μm	Many μm	Single emission lines
Modulation frequency	MHz	MHz	~10–100 Hz	~10 kHz
Spectral range	UV–MID IR	UV–MID IR	IR–MIR–FIR	UV

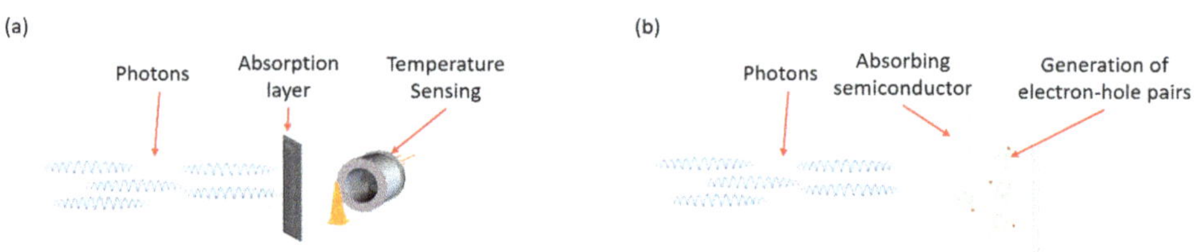

Figure 4.6. Schematic visulization of thermal light detectors and photonic light detectors. (a) Thermal sensors first convert the photon's energy to heat via a layer able to absorb photons, leading to a temperature increase, which is then detected. (b) In the case of photonic detectors a photon is converted into an eletron–hole pair and the associated change in the electronic state is recorded.

turn changes the electronic state of the material and consequently may be used to count the number of photons per second arriving at the detector area.

Photo resistivities consist of a semiconducting material and an electrode structure to determine its resistivity. Each absorbed photon will elevate an electron from the valence band to the conduction band, which leads to a change in electrical conductivity.

Photodiodes feature a p/n junction and the resulting space charge region serves two purposes: Firstly, its electric field accelerates created electrons and holes into opposite directions thus aiding the detection efficiency. Secondly, the speed of signal generation is higher.

It is important to note that one photon can create only one electron–hole pair regardless of its wavelength. As long as the photon energy is within the absorption spectrum of the semiconducting material each photon will cause the same signal, regardless of its frequency.

Thermal detectors

Various technologies to realize thermal detectors exist but their underlying operational principle is universal: photons are absorbed by a material and their energy is converted into heat via internal processes of the material. Ultimately, the energy of absorbed photons is converted into an increase in temperature. Measuring the temperature is the means to determine the light intensity. The most prominent examples of this group of detectors are pyroelectric detectors, bolometers, and thermopiles.

4.4.3 Interaction of photons and molecules on the optical path

As light travels through the optical path, absorption of photons is possible and the likelihood of this process happening crucially depends on the frequency of the photons as well as the optical path length and the number density of molecules $n = \#\text{molecules/m}^3$. The Beer–Lambert–Bouger law (Swinehart 1962) is at the heart of all absorption spectroscopy setups and relates those quantities using a simple model. It is based on a few assumptions and may be easily derived.

For the model to hold, the number density n of the molecules of interest as well as their frequency and temperature-dependent absorption coefficient $\alpha_T(\upsilon)$ have to be small such that absorption of the initial light with intensity I_0 along a short optical path length dz is weak. Furthermore, the molecules have to be distributed homogeneously. In that case, the attenuated light intensity dI is proportional to the optical path length dz:

$$dI = I_0 \cdot \alpha(\upsilon) \cdot dz. \tag{4.5}$$

Solving this first order differential equation above leads to the Beer–Lambert–Bouguer law:

$$I(z) = I_0 \cdot e^{-\alpha(\upsilon)\cdot z}. \tag{4.6}$$

A simple picture to describe the strength of absorption of a single molecule is using the so-called absorption cross-section $\sigma_T(v)$. As the name suggests it has the unit m^2 and describes the probability of a photon hitting and being absorbed by a molecule as it travels through space. The area, or cross-section, is dependent on the frequency of the photon and the temperature of the molecules. Far off a resonance, the cross-section will be very close to 0 cm^2, and consequently, the photon will reach the detector unhindered. However, as the photon energy approaches the energy difference between the two energy states of an allowed transition, the cross-section increases up to a finite maximum value. Hence the probability of photon absorption increases and, consequently, the detected light intensity is attenuated. The second factor influencing the attenuation is the number of molecules the photon passes on its way to the detector, determined by the number density n. Consequently, the absorption coefficient reads:

$$\alpha(v) = n \cdot \sigma_T(v). \tag{4.7}$$

Looking at the units involved in this process, it becomes clear that the actual spatial width of the probing beam has no influence on the results. This purely depends on the optical path length, the absorption cross-section, and the number density. An expression for the absorption cross may be derived from the line shape function $\phi(v)$, which describes the transition probability as a function of the frequency, and the so-called line strength S_T, which describes how many molecules are actually in the ground state in the first place. Per definition, the integral of the line shape function over all frequencies is 1. The line strength depends on the temperature of the gas, since it influences the probability of a quantum level actually being occupied or not. Equation (4.6) then may be written as:

$$I(z) = I_0 \cdot e^{-n \cdot \sigma_T(v) \cdot z} = I_0 \cdot e^{-n \cdot S_T \cdot \phi(v) \cdot z}. \tag{4.8}$$

The absorption cross-section is therefore a function of the line shape and the line strength. In order to understand the reason for the line shape structure, the underlying processes provide some guidance. The most important processes that give rise to the respective functions are:

- Natural linewidth:

  The most fundamental influence on the spectral bandwidth of a transition is the line shape resulting from spontaneous emission. There is a finite chance of the excited state decaying at any given moment and the process is entirely random. However, depending on the transition, the average lifetime τ_{nl} varies and in any case the probability for the molecule to remain in the excited state decays exponentially in time. In a semi-classical model, the finite lifetime of the oscillating charge in the associated excited state can be described by a damped harmonic oscillator, where the oscillation frequency corresponds to the frequency of the spontaneously emitted photon. Because of the finite lifetime τ_{nl} and the associated uncertainty of the emission time, the frequency of the photon is also not precisely defined. The spectral distribution of spontaneous emission is given by a Fourier transformation of the temporal

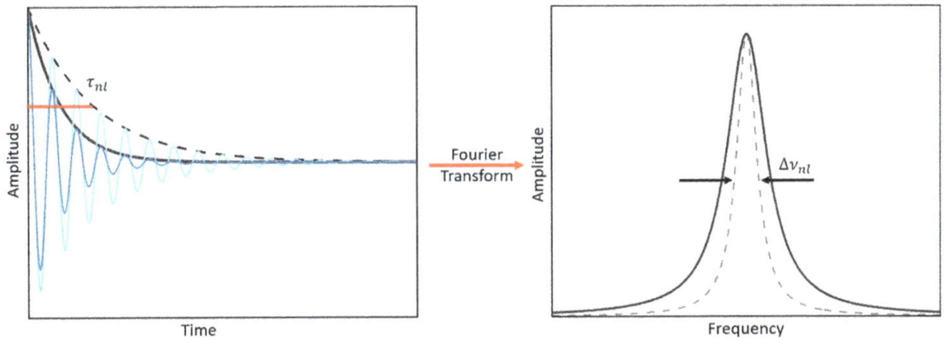

Figure 4.7. (a) The excited state lifetime decays exponentially featuring a finite lifetime τ of the excited state. (b) In consequence, a Fourier transform leads to the line shape associated with the transition from excited to ground state, which is of Lorentzian shape. In case collisions shorten the lifetime artificially, the associated bandwidth of the transition widens.

behaviour and leads to a Lorentzian line shape (Demtröder 2016) with a linewidth $\Delta\upsilon_{\text{nl}} \sim \frac{1}{\tau_{\text{nl}}}$. Since no other interactions or effects influence this behaviour it is named the natural linewidth. The process of spontaneous emission is pictured in figure 4.7, including the Fourier transform of the temporal behaviour.

- Pressure broadening:

  In most scenarios, many molecules are present at a given pressure p, which is equivalent to collisions between the molecules as well as with the container. Each collision may provoke a decay of an excited state of the molecule into the ground state and consequently lead to a pressure-dependent shorter lifetime. The more collisions occur, the shorter the excited state's lifetime. This in turn leads to a broadening of the linewidth in frequency space and since it depends on the number of collisions per unit time, it is called pressure broading or collision broadening.

- Doppler broadening:

  The molecules of a gas at constant, non-zero temperature randomly move with velocities v determined by the Maxwell–Boltzmann distribution (Müller-Kirsten 2013). The frequency of a photon interacting with those molecules υ is shifted with respect to the resonance frequency of non-moving molecules υ_0 depending on the relative movement of the molecule with the photon according to the Doppler shift (Walker *et al* 2014):

$$\upsilon = \upsilon_0\left(1 + \frac{v}{c_0}\right), \tag{4.9}$$

where c_0 is the speed of light. This means that the frequency shift of the gas is a function of the distribution of the velocities:

$$P(v)dv = \sqrt{\frac{m}{2\pi k_B T}} \cdot e^{-\frac{m\,v^2}{k_B T}} dv$$
$$\Longleftrightarrow \tag{4.10}$$
$$P_\upsilon(\upsilon)d\upsilon = \sqrt{\frac{mc_0}{2\pi k_B T\ \upsilon_0^2}} \cdot e^{-\frac{m\,c_0^2\,(\upsilon-\upsilon_0)^2}{2k_B T\ \upsilon_0^2}} d\upsilon,$$

which leads to a temperature-dependent Gaussian distribution of the resonance frequency of the gas molecules with a width $\sigma_\upsilon = \sqrt{\frac{k_B\,T}{m\,c_0^2}}\ \upsilon_0$.

The most important line broadening mechanisms for environmental gas sensing are Doppler and pressure broadening, which depend on temperature and pressure, respectively.

Depending on the relative influence of both processes the line shape is governed by Lorentzian shape or Gaussian shape. If both effects are about equally influential, then a convolution of Lorentzian and Gaussian function, the Voigt profile, is best used to model the line shape. Even then, the contribution of Doppler and pressure broadening may be distinguised because of the behaviour of the two classes of functions. At equal width, the Gaussian part of a Voigt spectral distribution will be dominant in the Kernel, while the Lorentzian shape will dominate the wings far off the resonance frequency.

The line strength S_T is mainly a function of the probability of connecting two internal energy levels by means of electromagnetic radiation and the probability of the ground state being populated. Since only dipole-allowed transitions have been discussed in this short introduction, this probability is determined by the electric dipole moment operator connecting those two states. The values and temperature dependence may be found in the HITRAN database (Gordon *et al* 2022).

4.5 Non-dispersive infrared absorption spectroscopy (NDIR)

The name of this technique historically derives from the fact that no disperive elements are required in this type of setup and that typically spectral features in the infrared range are used to measure the number density of molecules. This does not mean that the setup may not be used in other spectral regions. Still, the large infrared absorption cross-sections of vibrational transitions enables the construction of sensitive devices (Dinh *et al* 2016, Hodgkinson and Tatam 2013). A review and book for further reading on implementation and theory of NDIR are summarized in table 4.4.

Usually, a light source whose spectral linewidth is many orders of magnitude larger than single absorption lines of the molecules to be setup is shown in figure 4.8 and entails the light source, a measurement volume where photons and molecules may interact and two detectors with spectral filters in front of them.

One spectral filter in the optical path is used to limit the recorded light power to those spectral regions, where the gas of interest possesses strong absorption lines and

Table 4.4. Further literature detailing the NDIR technique.

Authors	Title	References
Wong and Anderson	'Non-dispersive infrared gas measurement'	Wong and Anderson (2000)
Dinh, Choi, Son and Kim	'A review on non-dispersive infrared gas sensors: Improvement of sensor detection limit and interference correction'	Dinh *et al* (2016)

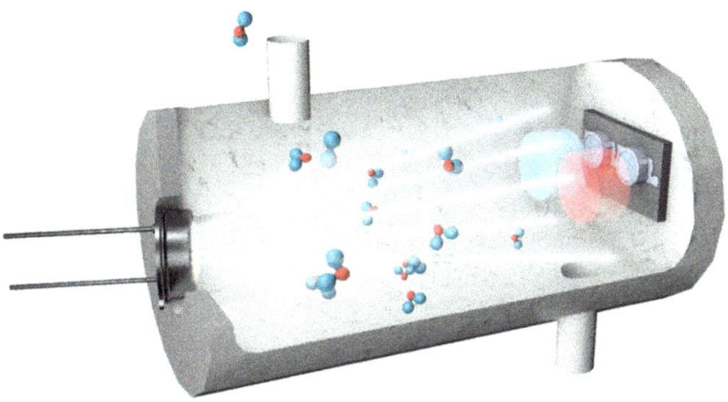

Figure 4.8. Schematic setup of an NDIR device: a broad band light source with a spectral emission function covering the absorption band of the target molecules is directed at two detectors. The spectral functions of the filters are chosen such that a reference and a measurement channel can be implemented.

limits the effects of cross-sensitivities. Additionally, to enable long-term stable operation many NDIR sensors are built with a reference channel. Their light path features a second spectral filter, whose function ideally excludes all regions with molecular absorption. This way fluctuations in the optical output power of the light source may be corrected for, assuming that the relative spectral emissivity remains unchanged. The signal of the NDIR measurement channel is a function of the respective spectral functions of all components, i.e. the emission spetrum of the light source $E(\omega)$, the spectral response function of the detector $D(\omega)$, the spectral transmission function of the filter $F(\omega)$, and the absorption features expressed by the absorption cross-section $\sigma(\omega)$ of the molecules. According to Beer–Lambert–Bouguer's law, the transmissivity $T(\omega, d) = e^{-n \cdot \sigma(\omega) \cdot d}$ of the gas sample is also dependent on the optical path length d along which light and gas molecules interact, and the number density of molecules n. Figure 4.9 schemetically depicts the various spectral functions, which are akin to those of typical NDIR setups. This highlights the necessity to use spectral filters, since without them the device would detect any molecular species within the spectral overlap region of light source and light detector. Even with the use of optical filters, an NDIR sensor will exhibit sensitivities towards gases within the measurement channel's filter transmission profile.

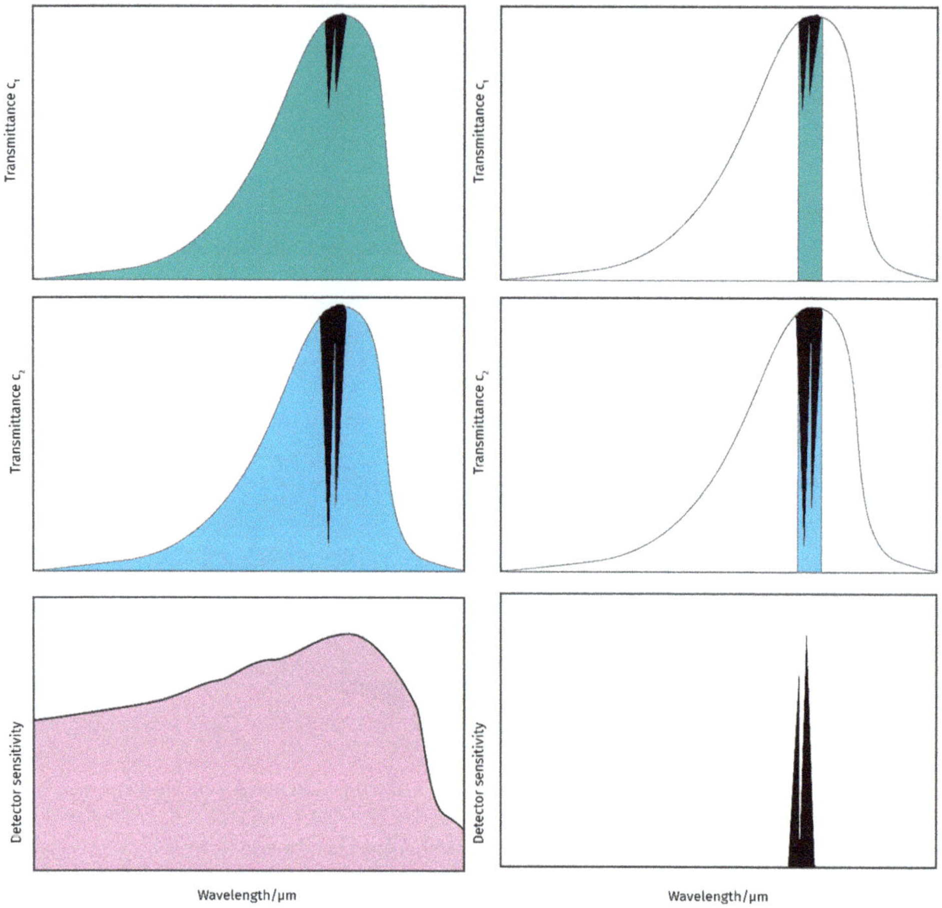

Figure 4.9. Spectral functions of the components involved in standard NDIR-based gas sensing. Cross-sensitivities occur whenever gases other than the target gas feature absorption lines within the filter bandwidth.

Additionally, the choice of the reference filter is of central importance, as it may also introduce cross-sensitivities.

The signal $S_{D,M}(d)$ generated by the measurement channel's light detector depends on the integral over the complete frequency spectrum:

$$S_{D,M}(d) \sim \int_0^\infty E(\omega) \cdot D(\omega) \cdot F_M(\omega) \cdot T(\omega, d)\, d\omega, \tag{4.11}$$

where $F_M(\omega)$ is the filter function of the measurement channel. Likewise, the signal $S_{D,R}(d)$ of the reference channel may be represented by substiting $F_M(\omega)$ with $F_R(\omega)$ in equation (4.6) if equal detectors are used. By dividing the two signals, variations in $E(\omega)$ and, to a lesser degree, $D(\omega)$ can be compensated for. Presuming that deteriorating optical components and thermal variations affected both channels equally, this method allows for building quite stable gas sensors. However, it is important to calibrate NDIR sensors to establish the relation between electrical output signal and number density and also since the reproducibility of individual components is not sufficient for achieving reliable concentration readings. The sensitivity of NDIR devices may be increased by increasing the optical path length. Consequently, the chosen optical path length very much depends on the intended application and expected concentration range. In general, the sensitivity function resulting from equation (4.11) does not result in a linear behaviour but oftentimes a linear approximation may be used for a limited range of number density values. Another relevant issue with NDIR sensors is their dependence on pressure: Users are often interested in the relative concentration of a specific gas species, i.e. a value stated in ppm or %. However, NDIR sensors do determine the number density, so a reasonably good approximation would be to use the ideal gas law to connect those two quantities. Oftentimes, this issue is disregarded altogether, because changes in the overall pressure are negligible and the calibration of the NDIR sensor establishes a direct correlation between desired output unit and electronic signal. However, it is an important point ot keep in mind should NDIR devices be used in applications with considerable pressure changes.

Selectivity evaluation and miniaturization

The simplicity of the setup is a main reason for its popularity. However, the price is paid in a lack of selectivity and sensitivity. In particular, all molecules that have absorption lines within the spectral profile of the filter of the measurement channel will contribute to the signal. This is especially significant in regions where many different gas species show absorption, e.g. the C–H stretch vibrations of hydrocarbon molecules in the 3 μm region of the infrared spectrum. On the other hand, if vibrational or electronic transition may be found that are not shared by many other gas species, NDIR systems may provide a selective alternative to other methods. The sensitivity function may be adjusted via the optical path length. Due to the broad bandwidth of the light sources, only multi-reflection cells are suitable for this task. At the same time, the required optical path lengths determine the potential for miniaturization.

Application scenarios

A standard application for NDIR sensors is carbon dioxide (CO_2) sensing. Since it is not very active chemically, most gas sensing technologies from the previous chapter are not a good option. Additionally, its strong absorption at 4.2 μm is not shared by many other, typically occuring molecules and the background concentration of CO_2 is already high as compared to other trace gases. If CO_2 is to be used as an indicator for air quality monitoring, then the concentration range form 400 ppm to about 5000 ppm is achievable with a few cm optical path length.

4.6 Tuneable diode laser absorption spectroscopy (TDLAS)

The so-called tuneable diode laser absorption spectroscopy (TDLAS) technique uses inverted roles of spectral functions of the main components as compared to NDIR. Herein, the light source features a spectral bandwidth much smaller than a single absorption line of the gas molecules. To gain information about the gas via absorption, the wavelength of the light source is altered and used to scan the absorption line. The tuning range is chosen such that spectral regions without any absorption and one or more individual absorption lines are recorded in each scan. The basic setup is depicted in figure 4.10.

As light source, single-mode laser diodes are regularly employed. In this context single-mode means that the laser diode emits only one single frequency with a spectral width mainly determined by the laser cavity. The central emission frequency of the laser is preliminarily determined via the spacing of the laser mirrors, which in turn depend on the temperature of the laser diode structure via the thermal expansion coefficient of the material. Consequently, the laser diode temperature T_{LD} is a means to adjust the diode laser wavelength $\lambda_{LD}(T_{LD})$ and while controlling the overall temperature is a possibility, it allows for a rather slow scan only. Much quicker thermal modulation may be achieved via the injected driving current i_{LD} of the laser diode $T_{LD}(i_{LD})$. Since the current induces Joule heating, this method is the prefered method in current TDLAS setups. The downside of this is the dependence of the optical output $I_0(i_{LD})$ power on the current, which increases linearly with i_{LD} in a first approximation. As a consequence, the scanning of the laser frequency is accompanied by a change in optical output power and this behaviour is shown in figure 4.11.

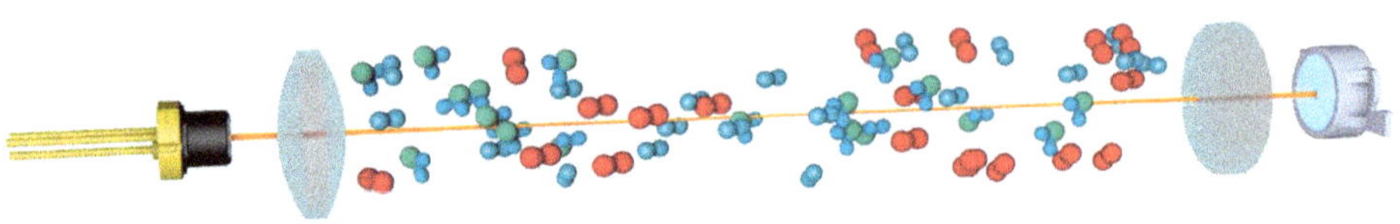

Figure 4.10. A typical setup of a tuneable diode laser absorptions spectroscopy setup consists of a collimated laser beam emitted by a laser diode, an optical path length, and a photodetector. The laser linewidth is on the order of a few MHz, while a single absorption feature is usually at least in the GHz range.

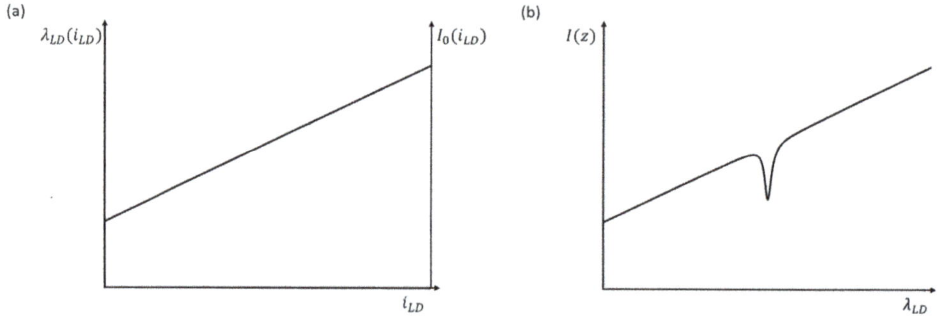

Figure 4.11. Exemplary behaviour of the correlation between diode laser current, output power, and laser frequency. (a) Changes in the laser's driving current lead to a change in emission wavelength and are accompanied by a change in output power. (b) The scan in wavelength is used to probe the absorption features of molecules. Due to the change in output power, the photodiode signal is composed of the absorption signal and changing output intensity I_0.

This way, alterations in the emitted power during operation are detected with every single scan and in this sense, every single measurement is self-referenced without the need for further components since in one scan, both absorption features and spectral regions without absorption are recorded.

The determination of the number density is now achieved via the Beer–Lambert–Bouguer law. Three parameters determine the absorption of light: the number density, the absorption cross-section, which is $\sigma_T(\nu) = S_T \cdot \phi(v)$ the product of the line shape function $\phi(v)$ and the temperature-dependent line strength S_T, and the optical path length *d*:

- The line shape is a normalized function, meaning that $\int_0^\infty \phi(v)dv = 1$.
- The line strength S_T is basically dependent on the occupation probability of the ground state that is involved in the transition probed, which is why it is temperature dependent and the probability of absorbing an adequate photon. Extensive documentation exists regarding values of S_T for most gas species and otherwise it will have to be determined experimentally (Gordon *et al* 2022).
- The optical path length is known, since the dimensions of the gas sensor are known.

By scanning over a complete absorption line, one effectively performes the integration of the line shape function, S_T is known from a look-up table and *z* is known from the design of the TDLAS setup. Consequently, the only unknown quantity in Beer–Lambert–Bouguer's law is the number density *n*. In order to obtain this from a measurement two steps are necessary:

1. Perform a baseline correction of the recorded signal to eliminate the influence of the intensity dependence on the diode laser current. While the baseline without absorption appears almost linear, it usually requires non-linear approaches to fit the corresonding curve and be able to compensate for light intensity changes by dividing the recorded signal by the baseline fit.
2. Once the pure absorption feature is obtained the total absorption has to be determined, i.e. the integration of the line shape function has to be performed. Usually, this is done based on an underlying model. If the observed absorption feature is mainly governed by pressure broadening, a Lorentzian function is used, whereas predominantly Doppler-broadened spectral lines may be fitted using a Gaussian profile.

The main advantage of this technique is the direct measurement of the number density based on the Beer–Lambert–Bouguer relation, which holds for low absorption levels. There are no futher calibration methods necessary, apart from determining the temperature of the gas sample. In this sense, each measurement of the absorption curve is a self-consistent, self-referenced determination of the number density. Nonetheless, oftentimes users like to be provided with results given in per cent (%), parts-per-million (ppm), etc. Using the ideal gas law, the relative concentration may be estimated via determining the pressure. It is important to bear in mind the limitations of the ideal gas law and usually corrections are necessary, depending on the environmental conditions.

Selectivity evaluation and miniaturization

Probing individual absorption lines of molecules enables specific detection and quantification of gas species. Even if spectral regions exhibit many absorption lines, identifying non-overlapping lines is often possible. However, the spectral linewidth is a function of the pressure and high-pressure scenarios typically lead to broad absorption features, which will ultimately hinder selectivity. To overcome this issue, controlling the pressure of the gas sample is an option that also enables isotope-selective gas sensing. Like in NDIR the miniaturization is dependent on the optical path length requirements. However, due to the narrow bandwidth of laser sources, the use of optical resonators is possible and several approaches for building miniaturized TDLAS-type instruments are ongoing.

Application scenarios

TDLAS is ideally suited for contactless sensing of arbitrary gas mixtures. This enables gas sensing at high temperatures and aggressive atmosphere, without the need to bring electronic and photonic equipment into direct contact with the sample. All that is needed is optical access and means to direct the laser beam to a photodetector. High temperature processes or toxic environments may be probed without the need for direct contact between instrumentation and the gas sample.

4.7 Photoacoustic spectroscopy

The photoacoustic effect has been known for more than 140 years. It has been described by various authors in parallel at the end of the 19th century and the fundamental process is based on the conversion of photon energy into heat via absorption of photons (Bell 1880, Mercadier 1881, Preece 1881, Röntgen 1881, Tyndall 1881). The preconditions for this have been discussed above. For all molecules where the relaxation after absorption is predominantly non-radiative, the photoacoustic effect tends to be strong, ultimately leading to an increase in temperature. A local increase in temperature in turn leads to an increase in pressure. Lastly, if the light absorption is periodically turned on or off, then this will lead to a periodic modulation of the pressure, which is a sound wave. The amplitude of that wave is proportional to the light absorption, making the sound volume propotional to the light intensity and the number of molecules excited. This approximation holds as long as the saturation intensity of the molecular transition is out of reach. The process of signal generation is depicted in figure 4.12.

The history of the use of the photoacoustic effect to determine gas concentrations is almost as long the knowledge of the effect itself. One particular use has been the improvement of NDIR sensors using the gas filling of a hermetically sealed cell as a spectral filter and detector of the light intensity. The sensing principle is still that of NDIR sensing but a photoacoustic cell is used to gauge the light absorption. This setup is beneficial for the selectivity, since only photons useful for the determination of the number density of molecules actually produce a signal in the first place. In fact, this has been the main reason for the development of early commercial setups,

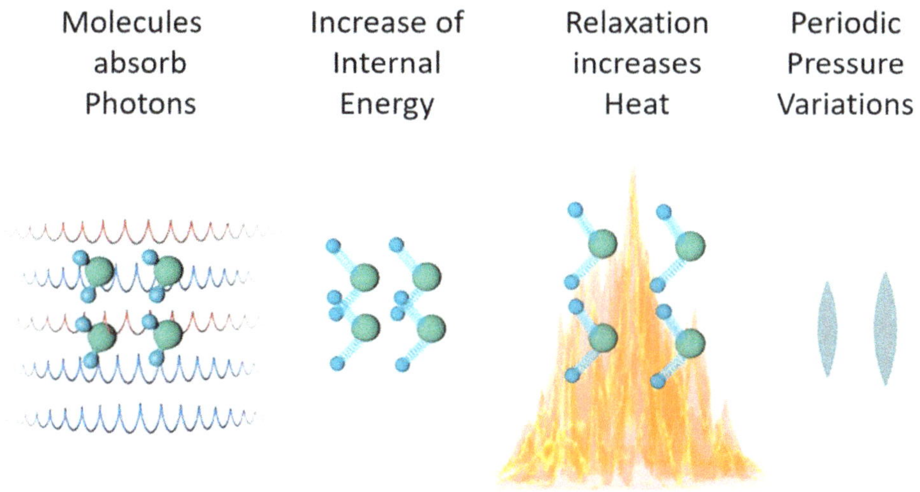

Figure 4.12. Schematic depiction of the photoacoustic effect: photon absorption leads to a local increase in heat, which is ultimately converted into increasing temperature leading to an increase in pressure. Consequently, a periodic modulation of light intensity leads to period pressure changes that may be recorded using a pressure transducer.

which has been sold under the name Ultrarotabsorptionsschreiber (URAS) for over 70 years. Instead of relying on a spectral filter to limit the spectral bandwidth of the light detected, the gas to be detected acts as a near ideal filter itself. A second benefit of using a photoacoustic detector is particularly important in the infrared spectral range: while semiconducting photodetectors are prone to thermal noise, photo-acoutic detectors employ a sound transducer to determine the light intensity, which has a far less pronounced temperature dependence. The URAS setup features a reference channel comprising a fixed number of absorbing molecules, oftentimes none, and a measurement channel. A thermal light source is intensity modulated at a fixed frequency by using a chopper and illuminates both channels and the detector. The elegance of this setup is twofold: Firstly, only light that is resonant to the absorption lines of the target gas will generate a pressure change in the detector chamber. Secondly, if both channels contain the same amount of target gas, then the signal generated by the membrane is zero, since changes in pressure are equal in phase and amplitude. Once absorption in the measurement channel occurs, the light intensity in that section of the detector will diminish and the membrane will record a sound signal.

Nowadays, this setup is still being used but an alternative method to utilize the photoacoustic effect has emerged. The setup is fundamentally different from the NDIR-type sensing approached and both approaches be clearly distinguished. If the photoacoutic signal is generated by exciting molecules in the direct vicinity of the sound transducer, then the method may be called direct photoacoustic spectroscopy. In case of NDIR-type setups, i.e. when the photoacoustic effect is used to gauge the light intensity in a hermetically sealed cell, the method is indirect.

4.7.1 Direct photoacoustic spectroscopy

The basic setup of direct photoacoustic spectroscopy is depicted in figure 4.13. It constitutes a light source that is adjusted such that it excites molecules next to a sound transducer. Via the photoacoustic effect a periodic pressure change is created and its amplitude depends on the light intensity and the number density of the gas. Importantly, all molecules with absorption lines within the spectral emission profile of the light source will generate a signal. To introduce selectivity in the setup two routes are taken:

1. Using a thermal emitter in combination with a spectral filter will limit the signal generation to those molecules that exhibit absorption features within the transmitted spectrum. Bascically, this will lead to a selecitivty performance comparable to standard NDIR sensors.
2. Employing a single-mode laser source leads to a selectivty comparable to TDLAS systems and in fact the procedure to quantify gas molecules is akin to the TDLAS technique.

Importantly, all direct photoacoustic spectroscopy methods have to determine the light intensity precisely in order to correctly correlate signal strength with number density of the gas molecules. Most actual setups are more complex in order to improve the signal-to-noise ratio. Engineering solutions aim at improving the insulation from ambient noise via buffer volumes, enhance the acoustic signal by using resonators, employing transducers with narrow bandwidth, and using optical resonators to enhance the light intensity available for signal generation. Additionally, lock-in techniques are typically used. The individual solutions are mainly determined by the user's requirements in terms of sensitivity, selectivity, and system size.

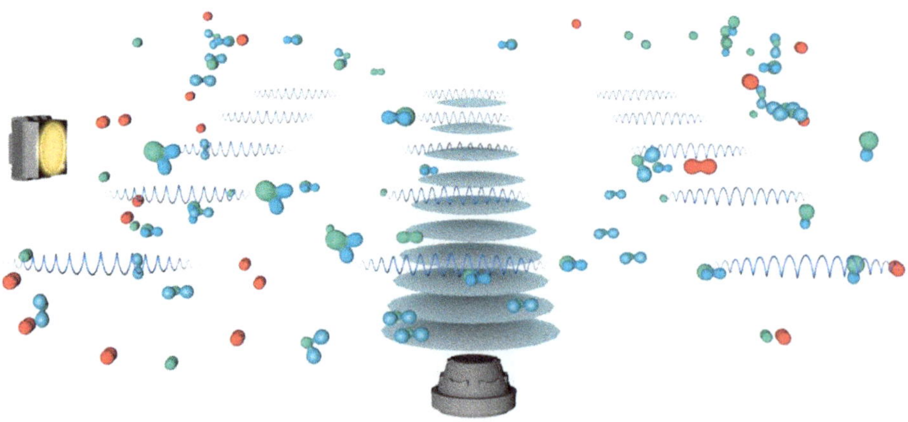

Figure 4.13. Schematic setup of a direct photoacoustic system: a modulated light excites source sound waves via the photoacoustic effect and may be recorded using a microphone, e.g. the acoustic wave's frequency is determined by the light modulation frequency, while the amplitude depends on the light intensity and the number density of the gas.

4.7.2 Indirect photoacoustic spectroscopy

The other basic, alternative setup to make use of the photoacoustic effect is similar to NDIR-based gas sensing. Instead of a photodiode and a spectral filter, a hermetically sealed cell filled with a gas and including a sound transducer is used. Equation (4.6) still holds but now the spectral response of the filter function $F(\omega)$ and the detector function $D(\omega)$ are determined by the gas filling and ideally in a photoacoustic detector the spectral response is equal to the absorption spectrum of the target gas. This can be achieved easily and to a high degree of accuracy by filling the photoacoustic cell with the target gas. Both the central absorption frequency of the individual transitions as well as their linewidth are mainly determined by temperature and pressure, so filling of the cell with 100% of the target gas at the same pressure as the deployment scenario of the gas sensor will generally yield good results in terms of selecitivity. The sensitivity of this setup may be adjusted via the optical path length, just as is the case in standard NDIR sensing.

One issue that arises with this simple setup is the implementation of a reference channel. If the original URAS setup is used, this is solved elegantly. However, the simple setup depicted in figure 4.14 is prone to changes in the emitted light intensity and deterioating optical components.

4.8 Fourier transform infrared spectroscopy (FTIR)

The absorption features of a gas are able to reveal the number of gas molecules in a sample as well as its pressure and temperature. Previously discussed methods have been mostly concerened with determining the number density of a single gas species. However, often gas matrices are highly complex and an analysis of the composition is required. One particularly elegant technology is based on recording the temporal

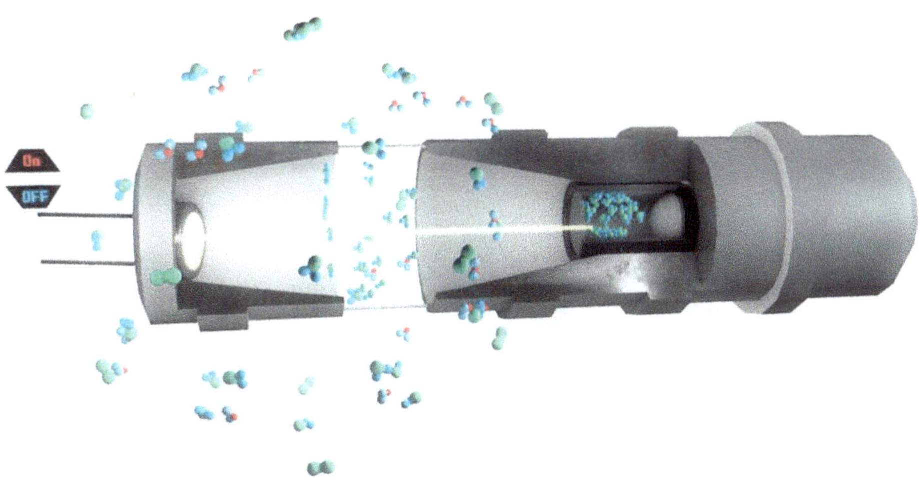

Figure 4.14. Schematic of an indirect photoacoustic setup. An intensity modulated light source excited acoustic signals recorded by sound transducer in a hermetically sealed cell. Any absorption originating from molecules with coinciding absorption lines will lead to a decline in photoacoustic signal strength.

coherence function of a broad band light source, whose emitted light has interacted with a gas sample and then calculate the Fourier transform of that function to obtain the frequency spectrum of both the light source and the gas (Smith 2011). To obtain the coherence function a Michelson interferometer is typically employed.

To gain a better understanding what a Michaelson interferometer does in this application it is depicted in figure 4.15. In the depicted configuration, the interferomenter is able to make the wave emitted by the light source interfere with itself but at different times: the light source's beam is split into two light waves with equal intensity and directed onto two mirrors, one of which may be moved. The light from both mirrors is retro-reflected and the beam splitter then recombines both light waves before the recombined wave is directed onto a photodiode.

By changing the relative travel length of light in both arms of the interferometer, a temporal difference is introduced. This means that by adjusting the optical path length difference, the light wave can be compared to itself via interference with itself. This way, the autocorrelation function, i.e. the temporal coherence function of the light source may be obtained and the autocorrelation time Δt is directly related to the displacement Δx:

$$\Delta t = c\ \Delta x, \tag{4.12}$$

via the speed of light c, i.e. a constant.

In the case of a monochromatic light source, the interference pattern recorded by the photodiode for different optical path length differences will yield a perfect sine wave. However, in reality the closest thing to a monochromatic wave is the light emitted by a laser. If one employes a narrow-band laser source, such as a He–Ne laser then the signal is descriptive since it resembles a light source that may be assumed as being close to monochromatic: one will obtain an interference signal with decreasing amplitude, given by the coherence length of the laser. Even assuming a frequency stabilized He–Ne laser as a light source, the interference pattern's amplitude, i.e. the contrast, will diminish with Δx, until ultimately no more interference will be detectable. Now, performing a Fourier transformation of the laser's autocorrelation function obtained via the Michelson interferometer will

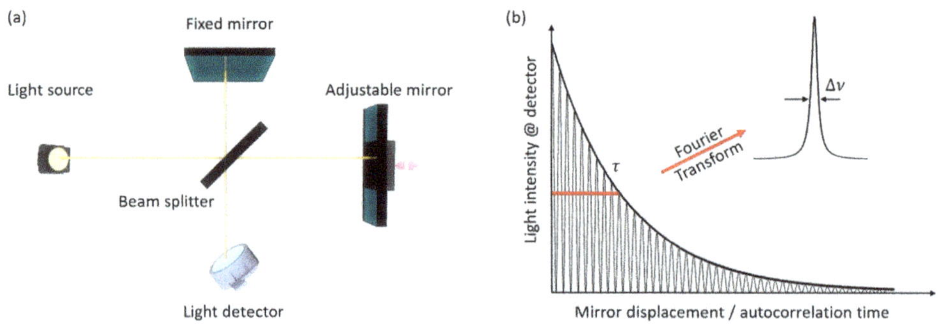

Figure 4.15. (a) The setup to obtain the light source's autocorrelation function requires a beam splitter that is use to divide the light wave into two equal parts. (b) The interference pattern obtained by adjusting the relative flight time of the two waves may be Fourier transformed and yields the spectral features of the light source.

provide the spectral emission profile of the laser. In the realm of FTIR, this determination of the coherence function is known as interferogram.

This highlights the usefulness of the Michelson interferometer for absorption spectroscopy: if one uses thermal light instead of a narrow-band laser the measurement will grant access to a spectrum spanning the complete infrared range. Using a computer to perform Fourier transformation of the so-called interferogram results in the frequency spectrum of that light source.

If one now places a gas sample between the light source and the detector, then those parts of the spectrum that exhibit absorption will be attenuated.

The typical setup of an FTIR is depicted in figure 4.16 and the central building block is a Michaelson interferometer. To prevent dispersion effects in the measured signal, all optical components are realized in reflective configuration. Typically, a frequency stabilized He–Ne laser is used to precisely determine the optical path difference Δx via recording its interference pattern. After the light has passed the Michelson interferometer it is funneled through the sample volume before it is recorded with a broad band infrared detector.

The interferogram is usually Fourier transformed using a fast-Fourier transformation (FFT) algorithm to obtain the entire spectrum of light source and gas sample in one go. In order to correct for background effects, two measurements are typically performed, one with and one without the gas sample. By dividing the results of both measurements, the spectral function of the light source as well as absorption features along the optical path are corrected for.

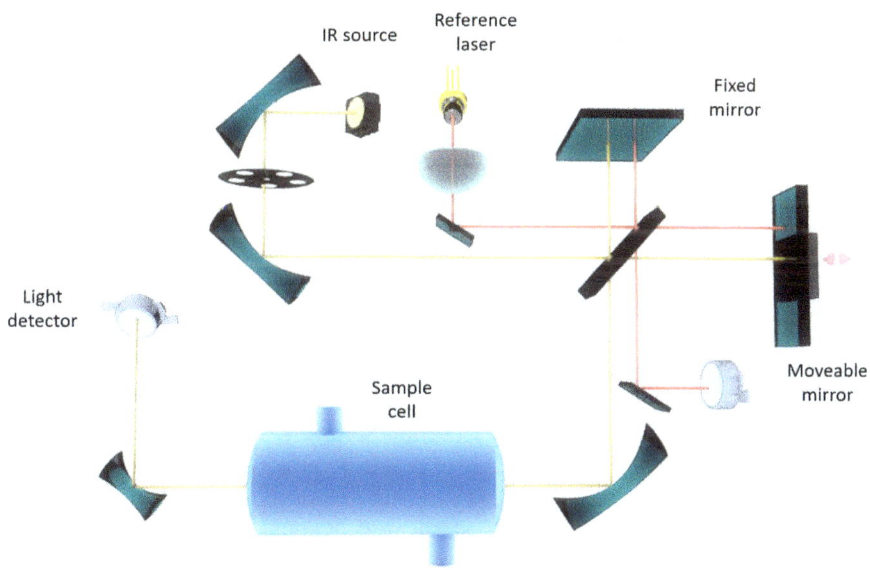

Figure 4.16. Schematic setup of an FTIR interferometer: the emission of a broad band thermal light source is directed towards a Michelson interferometer and subsequently through the gas sample before the intensity is recorded using a light detector. To determine the displacement, the Michelson interferemeter is used in parallel by a narrow-band laser source and the resulting intereference pattern allows for precise determination of the relative path lengths in both arms of the interferometer.

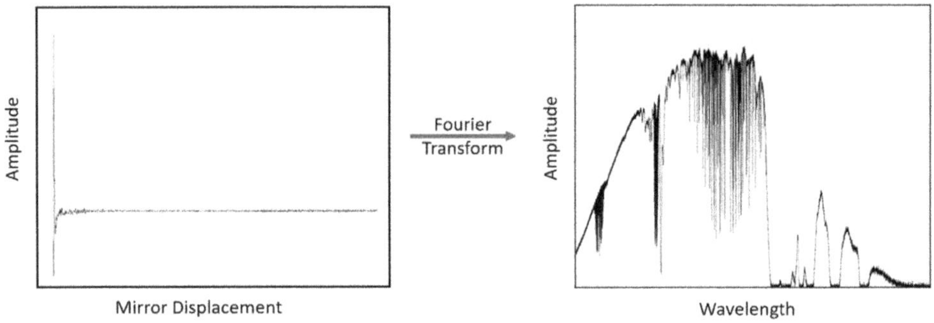

Figure 4.17. (Left) The interferogram contains the spectral information of the light source and absorption losses introduced by the gas sample. (Right) Using an FFT algorithm the gas sample's absorption features are extracted.

A sample measurement is depicted in figure 4.17 and highlights the advantage of FTIR over the other absorption spectroscopy techniques presented in this chapter: with a single measurement one can obtain the entire spectral absorption function of arbitrary gas matrices. The resolution of FTIR is fundamentally limited by the optical path length difference, as it determines the measurement time and in consequence the spectral resolution in frequency space. FTIR measurements are sensitive to any abrupt changes, e.g. by vibrations, since they introduce rapid changes in the interferogram, which translates to broad band spectral noise. Additionally, the discrete evaluation of the recorded interferogram by FFT algorithms may result in not resolving high-frequency features of the spectrum.

While FTIR is a powerful tool to investigate gas mixtures, it is also limited in terms of its potential for miniaturization and requires a high degree of vibrational damping. Without extensive adjustment deployment outside the laboratory environment is challenging. Furthermore, the quantification of individual gas species is difficult to obtain without calibration.

References

Aruldhas G 2014 *Molecular Structure and Spectroscopy* 2nd revised edn (New Delhi: PHI Learning)

Bell A G 1880 On the production and reproduction of sound by light *Am. J. Sci.* **20** 305–24

Demtröder W 2016 *Laser Spectroscopy* 5th edn (Berlin: Springer)

Demtröder W 2008 *Laser Spectroscopy* **vol 1** (Berlin: Springer)

Demtröder W 2015 *Laser Spectroscopy* **vol 2** (Berlin: Springer)

Dinh T-V, Choi I-Y, Son Y-S and Kim J-C 2016 A review on non-dispersive infrared gas sensors: improvement of sensor detection limit and interference correction *Sens. Actuators* B **231** 529–38

Gordon I E, Rothman L S, Hargreaves R J, Hashemi R, Karlovets E V, Skinner F M *et al* 2022 The HITRAN2020 molecular spectroscopic database *J. Quant. Spectrosc. Radiat. Transfer* **277** 107949

Gray H B 1995 *Chemical Bonds. An Introduction to Atomic and Molecular Sturcture* (New York: Freeman)

Hearnshaw J B 2014 The analysis of starlight *Two Centuries of Astronomical Spectroscopy* 2nd edn (New York: Cambridge University Press)

Hearnshaw J B 1990 *The Analysis of Starlight. One Hundred and Fifty Years of Astronomical Spectroscopy* 1st edn (Cambridge: Cambridge University Press)
Hecht E 2017 *Optics* 5th edn (Boston: Pearson)
Hodgkinson J and Tatam R P 2013 Optical gas sensing: a review *Meas. Sci. Technol.* **24** 12004
James J 2009 *A Brief History of Spectroscopy* ed J James (Hg.) (Spectrograph Design Fundamentals: Knovel) pp 1–5
Lewis G N 1926 The conservation of photons *Nature* **118** 874–5
Mallick P K 2023 *Fundamentals of Molecular Spectroscopy* (Singapore: Springer Nature)
Mercadier E 1881 Sur la radiophonie *J. Phys. Theor. Appl.* **10** 53–68
Mills A A 1981 Newton's prisms and his experiments on the spectrum *Notes Rec. Roy. Soc. Lond.* **36** 13–36 http://jstor.org/stable/531655
Müller-Kirsten H J W 2013 *Basics of Statistical Physics* 2nd edn (Singapore: World Scientific)
Newton I 2012 *Opticks. Or, A Treatise of the Reflections, Refractions, Inflections and Colours of Light Based on the Fourth Edition* (Dover Books on Physics) (Mineola, NY: Dover)
Preece W H 1881 I. On the conversion of radiant energy into sonorous vibrations *Proc. R. Soc. Lond.* **31** 506–20
Quack M and Merkt (Hg.) F 2011 *Handbook of High-Resolution Spectroscopy* (Hoboken, NJ: Wiley)
Raman C V and Bhagavantam S 1931 Experimental proof of the spin of the photon *Indian J. Phys.* **6** 353–66
Raman C V and Bhagavantam S 1932 Experimental proof of the spin of the photon *Nature* **129** 22–3
Reddy K V 2009 *Symmetry and Spectroscopy of Molecules* Revised 2nd edn (New Delhi: New Age International (P))
Röntgen W C 1881 On tones produced by the intermittent irradiation of a gas *Lond. Edinb. Dubl. Phil. Mag. J. Sci.* **11** 308–11
Saleh B E A and Teich M C 2001 *Fundamentals of Photonics* (Wiley Series in Pure and Applied Optics) (Hoboken, NJ: Wiley-Interscience)
Smith B C 2011 *Fundamentals of Fourier Transform Infrared Spectroscopy* 2nd edn (Boca Raton, FL: CRC Press) https://ebookcentral.proquest.com/lib/kxp/detail.action?docID=681305
Svanberg S 2022 *Atomic and Molecular Spectroscopy. Basic Aspects and Practical Applications* (Graduate Texts in Physics) 5th edn (Cham: Springer International Publishing)
Swinehart D F 1962 The Beer–Lambert law *J. Chem. Educ.* **39** 333–5
Tyndall J 1881 III. Action of an intermittent beam of radiant heat upon gaseous matter *Proc. R. Soc. Lond.* **31** 307–17
Walker J, Halliday D and Resnick R 2014 *Fundamentals of Physics* 10th edn (Hoboken, NJ: Wiley)
Weiner J and Ho P-T 2007 *Light—Matter Interaction: Fundamentals and Applications* (New York: Wiley)
Wollaston W H 1802 A method of examining refractive and dispersive powers, by prismatic reflection *Philos. Trans. R. Soc. Lond.* **92** 365–80 http://jstor.org/stable/107124
Wong J Y and Anderson R L 2000 *Non-Dispersive Infrared Gas Measurement: International Frequency Sensor* (Association Publishing/Raamatukogu)

IOP Publishing

Gas Sensor Technologies for Environmental Sensing

Stefan Palzer

Chapter 5

Combined systems—connecting selective elements with sensitivie detectors

In many applications the quantification and identification of numerous different types of molecules is required. Complying with both requirements for complex gas matrices is often only possible by employing a two-step approach consiting of firstly separating the different types of gas molecules either spatially or temporally, and subsequently determining the number density of those molecular species. To achieve gas species separation, different properties of molecules may be exploited, which usually involves their mass or their chemical behaviour. Mass separation typically involves ionization of molecules and it is therefore convenient to achieve the quantification by means of a current measurement. The quantification of neutral molecules may be achieved by several of the techniques presented earlier. This chapter briefly introduces the common concepts of separation techniques and discusses the respective applications.

The detection and identification of multiple gas species of a sample is difficult to obtain with the techniques presented so far. A notable exception is Fourier-transform-infrared spectroscopy (FTIR), which may be used to identify multiple gas species simultaneously. It is, however, limited in terms of resolution, as it depends on the availability of strong absorption features in the infrared, and does not provide direct means of quantification of each gas species. Alternatively, TDLAS setups would require about one laser source per gas species, which would quickly evolve into a complex apparatus. Furthermore, the constituents of the gas matrix have to been known in advance in order to select the corresponding laser wavelengths.

For these reasons, a different set of apparatus aims to provide a high level of selectivity between molecular types while enabling detection limits in the part-per-billion (ppb) range, which is ultimately limited by the detector used for quantification. Consequently, the design relies on two distinct sections to first perform separation and subsequently quantification. The fundamental methods that are

doi:10.1088/978-0-7503-3159-3ch5

used for separation of the molecule species are commonly employed depending on the properties of the gas samples. That might be a function of the respective masses, or via their chemisorptive behaviour. While more techniques for gas separation exist, those two are currently the most widely used for those measurment tasks, where the composition of gas samples is to be elucidated.

An additional distinguishing feature is the charging of the analyte, i.e. whether or not the samples are ionized. The organization of this chapter is in line with the ionization state of the analyte. First, techniques working with ionized molecules are discussed. In the second part of this chapter techniques relying on separation via chemical behaviour are presented.

5.1 Separation techniques using ionized samples

Analyzers working with ionized molecules generate ions from neutral samples and perform the quantification of molecules via charge counting. These common steps are complemented by several separation techniques, which lend the name to the respective approach. The fundamental process is depicted in figure 5.1 and the first and last step are discussed first, before the separation methods are presented in the respective sub-chapters.

To achieve reliable results, the generated ions of the sample gas need to be equally charged, which means that the objective is to achieve singly ionized molecules. The means to achieve electrically charging the molecules are diverse and table 5.1 gives a short overview of some of the possibilities to ionize gaseous samples. The choice of ionization technology depends on the gas matrix to be analyzed and the associated properties of the gas molecules. Special attention has to be paid to choose methods that do not alter the molecules, i.e. the ionization method should not make the molecule come undone. To this end, the different ionization approaches are classified as either 'hard' or 'soft' methods, depending on the energy deposited in the molecules. Several review publications are available on this topic (Awad *et al* 2015, Vikse and Scott McIndoe 2018) and here a brief overview is presented only. In general, hard ionization methods apply energies that may exceed some bonding strengths in the molecule, thus leading to splitting of the molecule. On the other hand, soft ionization methods may no be applicable always.

Ideally, this first step produces singly ionized molecules that are then funneled to the respective separation setups. Afterwards, the ions are usually counted via the

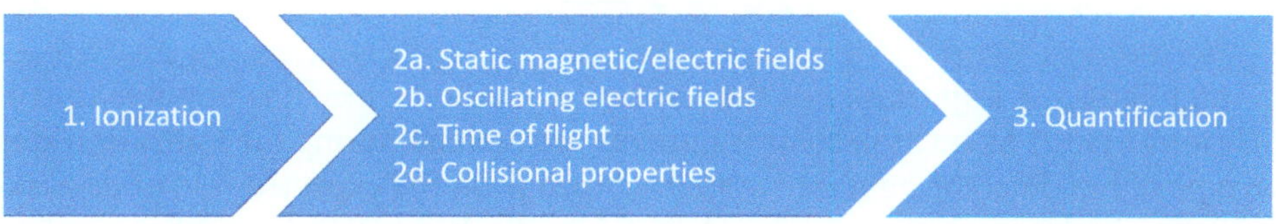

Figure 5.1. Techniques relying on the ionization of molecules and the subsequent separation of molecules according to their properties. While modern day equipment is highly complex, powerful, and usually makes use of a combination of effects, three basic, common types of separation are discussed here.

Table 5.1. Some of the methods that are used to generate ions are summarized here. The ionization method may be grouped into soft ionization techniques and hard ionization techniques.

Ionization method	Description	Further reading
Hard—electron impact ionization (EI)	Electrons are accelerated in a static electric field and used to bombard the gas molecules. Upon impact with the gas molecule part of the kinetic electron energy is converted to excite the molecule and promote it to an ionized state.	von Tilmann and Dunn (1985)
Soft—photoionization (PI)	High-energy photons are absorbed by the gas molecules and their energy suffieces to separate an electron from the molecule. This technique requires the photons to have more energy than the ionization energy of the molecules to be analyzed.	Zimmermann and Hanley (2021)
Soft—chemical ionization (CI)	This two-step method uses EI first to generate ions of reagent gas molecules, which subsequently interact with the gas molecules to be analyzed.	Harrison (2018)
Hard—inductively coupled Plasma (ICP)	A coil is used to produce an alternating electromagnetic field that heats up the gas inductively enough to create a plasma and hence a sufficient number of ions.	Houk *et al* (1980)
Soft—matrix-assisted laser ionization (MALDI)	A high-power laser is used to ionize samples embedded in matrix material.	Dreisewerd *et al* (2003)

resulting current using an electron multiplier (EM) (Allen 1947), a Faraday cup (FC) (Brown and Tautfest 1956), or a scintillator (Daly 1960). The different setups are depicted in figure 5.2 in a simplified version.

The transducing principle mostly relies on measuring the current, i.e. the number of charges that pass per time, very much similar to the way ions are detected in flame ionization detectors (FIDs) and phototionization detectors (PIDs). The direct relationship between the recorded current i and the number of ionized molecules N_{ion} provides reliable means for quantification:

$$i = \frac{Q}{t} = \frac{N_{\text{ion}}}{t}. \tag{5.1}$$

This highlights the importance of the ionization method, since it is possible to count single ions as long as no molecular fragments are created during the ionization stage. The type of ion detector built into systems depends on the expected rate of detection. While the high amplifications in electron multipliers (EMs) enables single-ion counting with high resolution, the dynamic range is limited and detecting high number densities of molecules may not be possible.

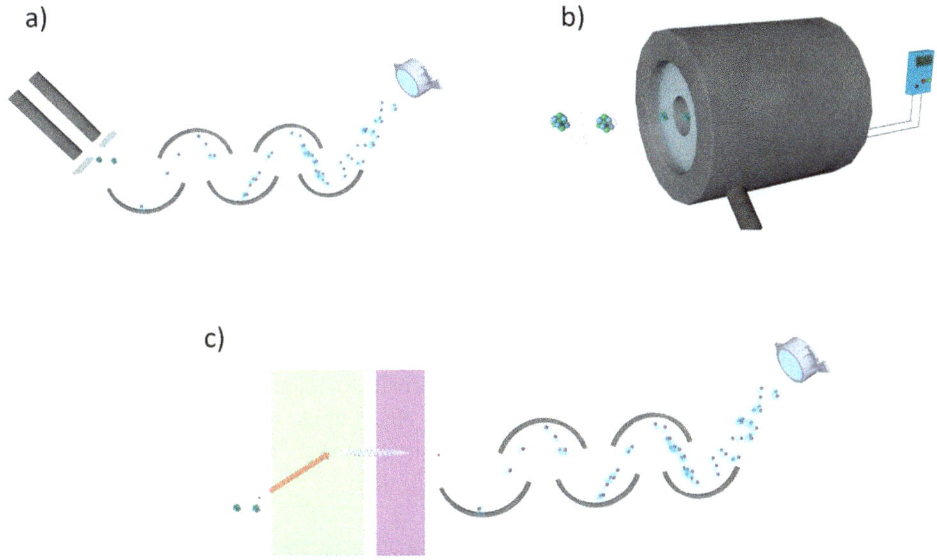

Figure 5.2. The three most common ion detectors used in complex gas sensing appartus rely on determining the current resulting from an ion flow. (a) An electron multiplier accelerates electrons at each amplification step thus multiplying the available charges in every step. (b) FCs are simple setups where the ions ultimately create a voltage that is detected. (c) Scintillator setups rely on the conversion of ions to photons, which in turn are detected using a photomultiplier.

5.1.1 Mass spetrometers

Mass is one of the fundamental properties of all types of molecules and many techniques aim to determine the mass of a gas species by determining the charge-to-mass ratio q/m (Griffiths 2008). These methods are basically counting the combined number of protons and neutrons in a molecule, since the electron mass can usually be disregarded. Therefore, the capability to identify molecule types in a complex gas matrix are closely related to the dynamic range of those apparatuses as well as their resolution, i.e. the smallest differences in molecular mass that the device can separate reliably. In general, mass spectrometers rely on employing the interaction between a charge and either static or dynamic electric or magnetic field, or a combination thereof to achieve a mass separation. Table 5.2 provides an overview for further reading on the different technologies that leverage the mass of ions for their separation, which is achieved either via recording different times of arrival or arrival points.

A. *Sector mass spectrometry*

A charged particle moving in a magnetic field will experience the so-called Lorentz force $\overrightarrow{F_{\text{Lor}}}$ perpendicular to both, the velocity vector $\vec{v}$ of the ion and the magnetic field vector $\vec{B}$, as well as experience an acceleration parallel to the electric field vector $\vec{E}$:

$$\overrightarrow{F_{\text{Lor}}} = q \cdot (\vec{E} + \vec{v} \times \vec{B}). \tag{5.2}$$

Table 5.2. Introductory review articles for separation methods based on ions. The different approaches are presented in the respective section in a concise fashion.

Section	Technology	Separation method	Title	Further reading
A	Sector mass spectrometry	Spatial	'An introduction to ion optics for the mass spectrograph'	Burgoyne and Hieftje (1996)
B	Time of flight	Temporal	'Time-of-flight mass spectrometry: Introduction to the basics'	Boesl (2017)
C	Quadruple mass spectroscopy	Spatial	'Quadrupole mass analyzers: Performance, design and some recent applications'	Dawson (1986)
	Ion trap	Spatial	'Ion traps in modern mass spectrometry'	Nolting *et al* (2019)

In so-called sector mass spectrometers, different regions of the instrument will make use of either the magnetic or electric component of the Lorentz force. This means that a well-defined sector of the instrument features either a constant magnet or electric field, while the respective other field is zero. Different sectors may then be combined to improve the overall performance of the apparatus. For example, a classic setup consists of an electric sector, followed by a magnetic sector to make up a so-called two-sector mass spectrometer.

As evident from fundamental classical mechanics, the acceleration resulting from the Lorentz force is dependent on the mass and this results in different trajectories of movement. In the so-called sector mass spectrometry this is used to determine the number of ions that have travelled along a certain trajectory, which is dependent on the molecule's mass. Depending on the design of the apparatus the resolution in terms of mass separation capabilitiy may be tuned. The limit of detection depends on the detector and is typically in the ppb range, where the concentration of a molecular species is directly proportional to the current detected.

The mass separation in a constant magnetic and absent electric field is achieved via the different trajectories of ions of different mass but equal velocity. This becomes clear by equating the Lorentz force with the centrifugal force $F_Z = m\,\frac{v^2}{r}$ and calculating the radius r of the resulting circular motion:

$$\overrightarrow{F_{\mathrm{Lor}}} = q \cdot v \cdot B = m \cdot \frac{v^2}{r} = \overrightarrow{F_Z} \\ \Leftrightarrow \\ r = \frac{m \cdot v}{q \cdot B}. \tag{5.3}$$

This means that the radius of the ion trajectory depends on the momentum of the ion. Because the radius is therefore dependent on the incoming velocity of the ion, a velocity filter may be applied first in order to achieve a narrowing of the ions' velocity distribution, since the velocity distribution limits the mass resolution of this setup. A schematic graph of the working principle for separating ions by their mass via a static magnetic field is depicted in figure 5.1. The first such instrument was desribed more than 100 years ago (Dempster 1917) and nowadays sector mass spectrometers are still among the most powerful devices in mass spectrometry. A sector mass spectrometer may be operated continuously and the schematic setup is depicted in figure 5.3. Typically, mass spectrometry is able to separate different isotopes of the same gas. However, it is important to keep in mind that this technique detects only the mass and not the structure of a molecule.

Instead of providing a multitude of detectors to register ions with different radii, the magnetic field is usually scanned, such that ions with different masses reach a fixed detector for different magnet field strengths, i.e. different masses.

The Lorentz force resulting from a purely electric field may also be used in a similar way by using a curved capacitor applying an electric field to force ions on a circular path:

$$\overrightarrow{F_{\text{Lor}}} = q \cdot E = m \cdot \frac{v^2}{r} = F_Z \\ \Leftrightarrow \\ r = \frac{m \cdot v^2}{q \cdot E}. \tag{5.4}$$

This is how the so-called electric sector is designed. In this case, the trajectory depends on the kinetic energy of the ion, as opposed to the momentum of the ion in the case of the magnetic sector.

This opens up the possibility to correct for varying velocities at the input of magnetic sector. Since the magnetic sector diverts ions according to their momentum and the electric sector according to their kinetic energy, both sectors may be combined to correct for the respective broadness of the velocity or kinetic energy

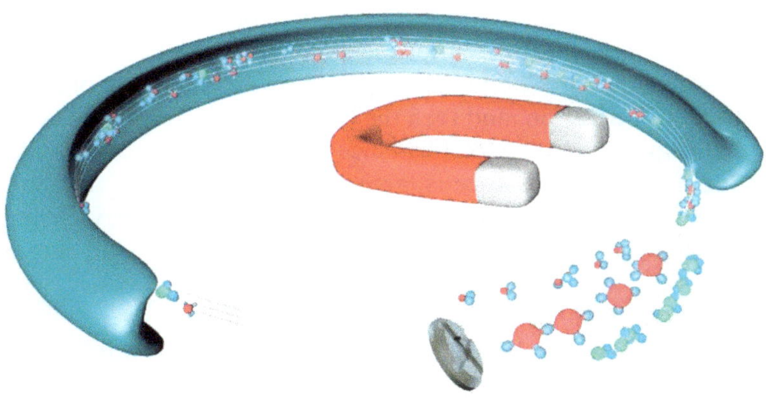

Figure 5.3. Ions of equal velocity are loaded into a sector with a constant magnetic field. The radius of circular shaped trajectory that the ion adopts due to the Lorentz force they experience is dependent on their mass.

distribution. For this reason, at least two sectors made up of an electric and magnetic sector are usually combined. Historically, a number of different setups have been proposed and used and nowadays a more complex system provides ever improved performance (Grayson 2002).

B. *Time-of-flight mass spectrometry*

An alternative way for mass separation is based on acceleration of an ionized gas sample by a static electric field that is generated by a plate capacitor. In this simple case, ionized molecules are accelerated in a static magnetic field, where the field E is a function of the potential difference between the two plates ΔV, and the distance d between the plates:

$$E = \frac{\Delta V}{d}. \tag{5.5}$$

Using equation (5.1) it becomes clear that the potential energy E_{pot} of singly ionized molecules in such a field is:

$$E_{\text{pot}} = q \cdot E \cdot d = q \cdot \Delta V. \tag{5.6}$$

In time-of-flight mass spectrometry this is used to provide the generated ions with the same kinetic enery, i.e.:

$$E_{\text{kin}} = \frac{m}{2}v^2 = q \cdot \Delta V = E_{\text{pot}}. \tag{5.7}$$

If each ion is singly ionized then the acceleration of individual ions resulting from the static electric field will depend on its mass only. In this case, the velocity $v_{\text{f}} = \sqrt{\frac{2 \cdot q \cdot \Delta V}{m}}$ upon passing the capacitor will show a mass-dependent value, while the kinetic energy is equal for all the ionized molecules. The generated ions are then allowed to travel a distance d prior to reaching the detector. The time of travel $t_{\text{d}} = v_{\text{f}} \cdot d$ is entirely dependent on final velocity after acceleration for a fixed drift distance d, hence masses may be determined according to the respective time of arrival at the detector. The pre-requisite for this to work is to have a well-defined ion pulse to begin with. Time-of-flight spectrometers therefore can only be operated in a pulsed mode, where in a first step ions are generated, then a pulse of ions is released via a time limited acceleration, and then the time of arrival for the different masses is registered. An improvement in performance may be achieved by building larger devices, such that the drift distance is larger and, consequently, the time of arrival increases. Folded drift paths using electrostatic ion mirrors may be implemented such as to increase the length without increasing the device's footprint figure 5.4.

C. *Quadrupole mass spectrometer*

The inertia of molecules may be leveraged in a less straightforward way to separate masses. The basic idea is to use rapidly oscillating electric fields to generate a time-averaged potential for charged particles. The Maxwell equations do not allow for a global minimum of the electric field, which is why this trick is necessary. However,

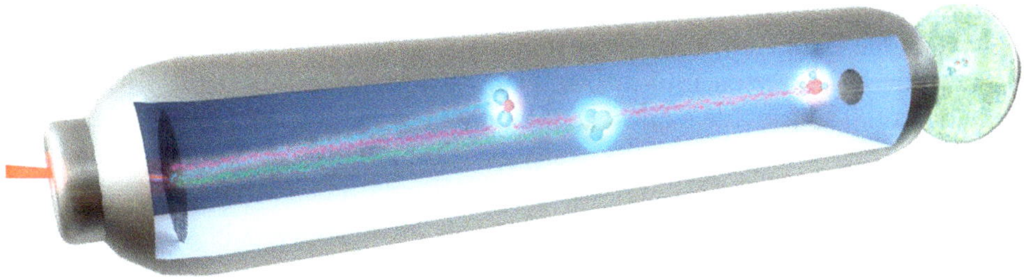

Figure 5.4. After an ion pulse has been generated, the ions travel along a distance *d* with the velocity acquired by the initial acceleration. The time of arrival of ion pulses is recorded and the time of flight enables the determination of the mass. The schematic setup depicts a simplified version. Modern devices make use of sophisticated methods to enhance the travel distance and minimize differences in the initial kinetic enery of the ion pulse.

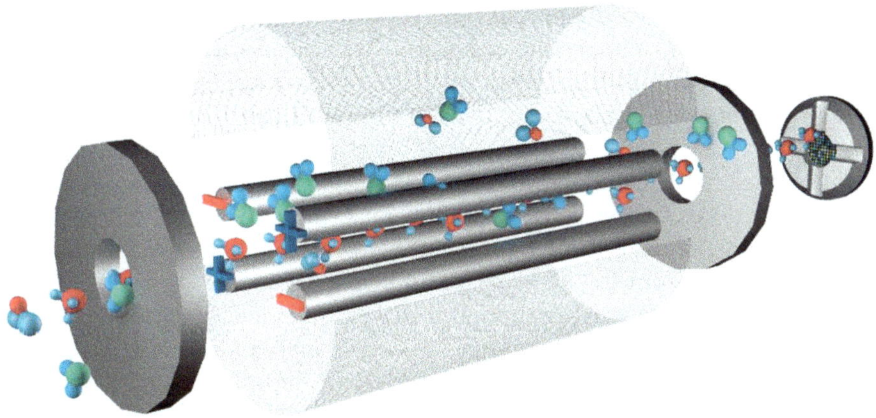

Figure 5.5. An oscillating quadrupole field may be used to construct an effective trap for ions.

an alternating electric field may be used to construct an effective way to provide stable trajectories for ions. The equations of motion for charged particles in such a field can be analyzed and lead to the so-called Mathieu equations (Ghosh 2007), whose solutions determine mass-dependent, stable regions for charged particles. Those regions may be tuned via constant electric offset fields or adjusting the oscillation frequency. This basic property enables the building of mass selective filters and an exemplary setup to highlight the idea is depicted in figure 5.5. The idea behind quadrupole mass spectrometers is to enable stable trajectories for only one specific mass at a given time and detecing how many ions make it through the electrode structure. By changing the operational parameters, one is able to scan the different ion masses present in a gas sample. The mass resolution of those setups is lower when compared to the alternative methods presented above. However, the potential for miniaturization is higher.

The mass filter setup may be adjusted to build ion traps, which are also known under the name Paul trap. Instead of sending a stream of ions along the electrodes, the ion guide is converted into an ion trap by adding end electrodes with a static electric potential to provide a lateral confinement of the ions as well. This way, long-term stable trajectories for all ion masses within the corresponding operational parameters is provided. Via changing the oscillation frequency, the trapping potential for different masses may be adjusted. Since changes in frequency can be applied with a high degree of precision, the ejection of ions with different masses may be triggered this way. That means that the operational protocol for mass spectrometry relies on filling the trap with a range of different masses and then tuning the trapping frequency to selectively eject different masses from the trap, which in turn are detected. As compared to a mass filter setup, the selectivity is improved by roughly one order of magnitude.

5.1.2 Ion mobility spectrometry

A technique akin to time-of-flight mass spectroscopy but based on a different type of separation mechanism, namely the collisional parameters between gases, is called ion mobility spectroscopy. The setup of those systems appears similar to time of flight spectrometers at first sight, but instead of operating in a vacuum, the ions are introduced into a constant flow of a neutral, inert gas flowing in opposing direction to the ions, which are accelerated in a constant electric field. The frequent collisions between ions and the counter-moving neutral ions cause an almost constant velocity of all types of ions, regardless of their weight. In the low field-strength domain the velocity is nearly constant for all types of ions and dependent on the field strength. However, the size of the ions is the main driver influencing the collisional cross-section σ_{Col} and consequently, the number of collisions per unit time. Along with the number density of the gas mixtures, this leads to an ion mobility K, which depends largely on the size of ions via the scattering cross-section. For small electric fields this can be approximated by the Mason–Champ equation (Mason and McDaniel 1988):

$$K = \frac{3}{16} \cdot \sqrt{\frac{2\pi}{\mu\, k_{\mathrm{B}}\, T}} \cdot \frac{q}{n \cdot \sigma_{\mathrm{Col}}}, \tag{5.8}$$

where $\mu = \frac{m \cdot M}{(m + M)}$ is the reduced mass of collisional partner, i.e. ions and neutral gas.

The collisional behaviour of the ions with the gas flow now causes different times of arrival at the detector, and the time ions drift through a neutral flow of gas enables the separation of molecule types according to their collisional properties. The time t it takes for an ion species to reach the ion detector may then be expressed as:

$$t = \frac{L}{v_{\mathrm{Drift}}} = \frac{L}{K \cdot E}, \tag{5.9}$$

where L is the length of the drift region. Lastly, ions have to be detected and this is usually done by recording the number of ions that pass per time.

5.2 Gas chromatography

Forces that may be exerted on neutral molecules by electromagnetic fields are typically weaker as compared to working with charged particles. Alternative separation approaches for neutral molecules may instead involve chemical interactions and one particularly important technique in this regard is so-called gas chromatography. The name gas chromatography, i.e. colour writing, originates from the early usage of the technique to separate pigments (Tswett 1906). One important implementation relies on using different adsorption properties of gas molecules with either a solid or liquid. The underlying idea is that the time a gas molecule is either adsorbed to a surface or absorbed by a liquid depends on the individual chemical properties of that gas species and the respective reaction partner. This creates the so-called stationary phase in chromatography, since the molecules are effectively immobilzed.

For the sake of simplicity and without limiting the general applicability of the description, the following discussion is limited to gas–solid chromatography, meaning that gas molecules may adsorb on a solid surface. Hence the process relies on adsorption, which has been discussed earlier in a different context and it immediately becomes clear that all processes in gas chromatography are temperature dependent and the behaviour of such systems is governed by the equilibirum state between adsorbed and free molecules. The denomination in the realm of gas chromatography is slightly different though, since it is concerned with the global behaviour of a gas species in a complex gas matrix. Therefore, the adsorbed molecules are referred to as the stationary phase and the free molecules as the mobile phase. At a fixed temperature a characteristic equilibrium between those phases exist, which depends on the chemical interaction between surface and gas molecules. This means that two phases co-exist and the ratio of the number densities between stationary phase n_s and mobile phase n_m is expressed as the temperature T and molecule species(s)-dependent, so-called distribution constant $K_C(T, s)$:

$$K_C(T, s) = \frac{n_s}{n_m}. \tag{5.10}$$

The basic working principle of gas chromatography now relies on injecting a gas sample into a probe volume and using a constant flow of an inert gas to move the molecules along in a capillary or packed column. The inert gas movement is not delayed in its movement by adsorption processes and typically gases like helium are employed. The flow rate of the so-called carrier gas hence determines the velocity of the mobile phase, since molecules are dragged along by collissions. Consequently, the time for the mobile phase to cross the column t_m is a function of the inert gas flow and the length l of the column, both of which are typically constants. The principle of gas species separation via gas chromatography is depicted in figure 5.6. A gas sample is introduced into the system at $t_0 = 0$ s and exposed to a constant gas flow of the carrier gas. Depending on K_C, the retention times vary and the different species are separated via their arrival time at the detector.

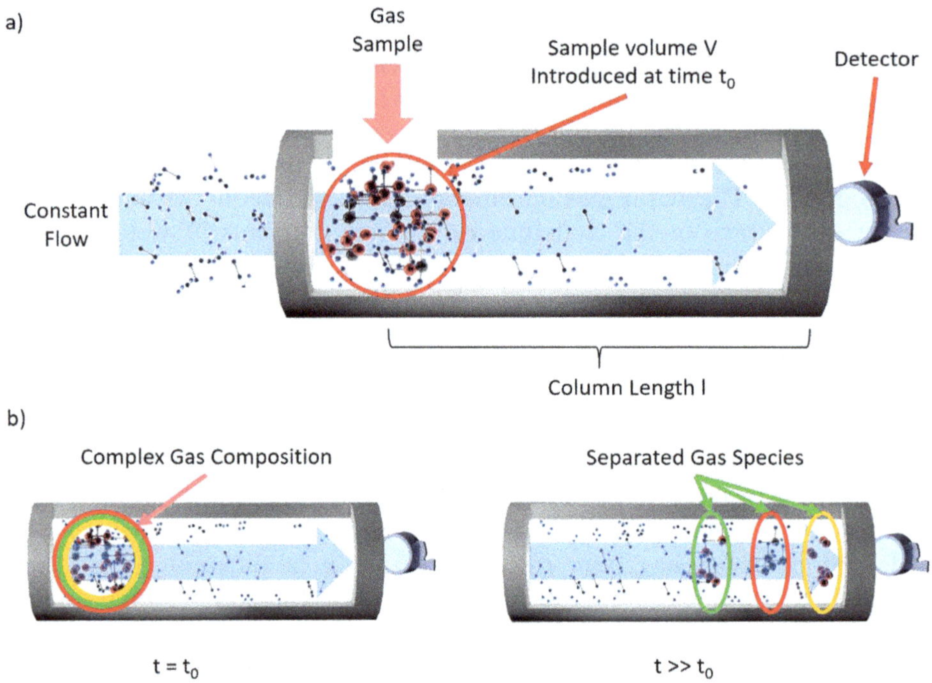

Figure 5.6. A schematic overview of a gas chromatographic system and the main parameters determining its behaviour. (a) The setup involves a constant flow of carrier gas and a mechanism to introduce a well-defined volume of a gas sample at a well-defined time. (b) After a retention time, the molecules are detected using a sensitive detector.

However, the stationary phase slows down the movement of a molecular species s, since stationary phase and mobile phase for a given molecular species are in equilibrium at all times. In particular, the time in stationary phase t_R' delays the time of arrival resulting in the molecular specific so-called retention time t_R:

$$t_R = t_m + t_R'. \tag{5.11}$$

The ratio of the number of molecules in the respective phase $\frac{\#s}{\#m}$ is equal to the ratio of time in the respective phase $k = \frac{t_R'}{t_m}$, which is called the retention factor and leads to expressing the distribution constant in terms of the involved times and volumes of the chromatography system:

$$K_C(T, s) = \frac{n_s}{n_m} = \frac{\#_s \cdot V_m}{\#_m \cdot V_s} = \frac{t_R' \cdot V_m}{t_m \cdot V_s} = k \cdot \beta. \tag{5.12}$$

The ratio of the volumes occupied by molecules in the mobile and solid phase V_m and V_s is denoted the phase ratio β. Since K_C depends on the molecule species, the initially mixed gas matrix will be separated on its way through the column according to distribution constant, i.e. different average times attached to the wall. In effect, this leads to the so-called retention time being different for different gas species, i.e. some gas types make it through the capillary quicker than others.

Two main types of basic column setups are usually employed, i.e. capilary and so-called packed columns. In both cases the time it takes for a gas species to pass the column depends on its adsorption properties and the resulting retention time to the detector upon injection of a gas sample is recorded. Consequently, the length and making of column as well as its temperature are the determining factors for separation and selectivity. A high surface-to-volume ratio is beneficial for achieving relatively strong adsorption influence, which explains the two main types of column designs. For the case of a capillary system, the phase ratio may be estimated easily, assuming an adsorption layer of thickness d_s much smaller than a capillary of radius r_c and a molecular distribution width w_s:

$$\beta = \frac{V_m}{V_s} = \frac{\pi \cdot (r_c - d_s)^2 \cdot w_s}{2\pi \cdot r \cdot d_s \cdot w_s} = \frac{(r_c - d_s)^2}{2 \cdot r \cdot d_s} \approx \frac{r_c}{2d_s}. \tag{5.13}$$

The phase ratio is an important design parameter in chromatographic devices and for a packed column their calculation is considerably more complex. Figure 5.7 illustrates the working principle of a gas chromatography system and the parameters discussed thus far.

After passing through the capillary, the molecules are detected via an unspecific yet sensitive detector, such as an FID, a PID, or a thermal conductivity detector. The latter determines the sensitivity of the system and via calibration of the generated signal a number density may be calculated. The respective choice for the detector technology depends on the type of molecule that is expected in the gas samples. The level of selectivity of gas chromatographs is mainly determined by the separation capability of the column. When operating a gas chromatography system, the method and volume of introducing the gas sample are important factors.

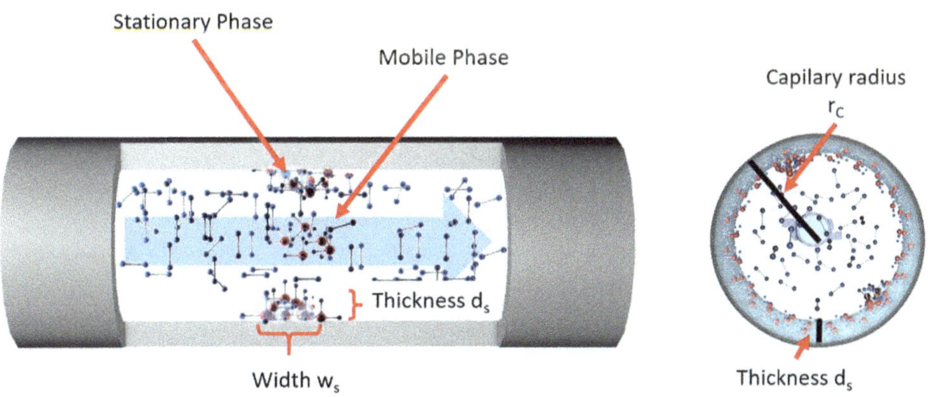

Figure 5.7. The inert carrier moves along the gas species, which are distributed among a stationary phase and a mobile phase in an equilibrium condition at all times. (Right) Cross-sections of the mobile and stationary phase at a fixed time of capillary with radius r_c. The stationary phase features a thickness and width occupying a volume with a number density n_s. The mobile phase is in equilibrium with the stationary phase at all times and occupes a volume V_m with a number density n_m.

References

Allen J S 1947 An improved electron multiplier particle counter *Rev. Sci. Instrum.* **18** 739–49

Awad H, Khamis M M and El-Aneed A 2015 Mass spectrometry, review of the basics: ionization *Appl. Spectrosc. Rev.* **50** 158–75

Boesl U 2017 Time-of-flight mass spectrometry: introduction to the basics *Mass Spectrom. Rev.* **36** 86–109

Brown K L and Tautfest G W 1956 Faraday-cup monitors for high-energy electron beams *Rev. Sci. Instrum.* **27** 696–702

Burgoyne T W and Hieftje G M 1996 An introduction to ion optics for the mass spectrograph *Mass Spectrom. Rev.* **15** 241–59

Daly N R 1960 Scintillation type mass spectrometer ion detector *Rev. Sci. Instrum.* **31** 264–7

Dawson P H 1986 Quadrupole mass analyzers: performance, design and some recent applications *Mass Spectrom. Rev.* **5** 1–37

Dempster A J 1917 A new method of positive ray analysis *Phys. Rev.* **11** 316–25

Dreisewerd K, Berkenkamp S, Leisner A, Rohlfing A and Menzel C 2003 Fundamentals of matrix-assisted laser desorption/ionization mass spectrometry with pulsed infrared lasers *Int. J. Mass Spectrom.* **226** 189–209

von Tilmann D M and Dunn G H 1985 *Electronic Impact Ionization* ed D von Tilmann and H D Märk und Gordon (Wien: Springer)

Ghosh P K 2007 *Ion Traps* (The International Series of Monographs on Physics, 90) (Oxford: Clarendon)

Grayson (Hg.) M A 2002 *Measuring Mass: From Positive Rays to Proteins* (Philadelphia, PA: Chemical Heritage Press) http://loc.gov/catdir/enhancements/fy0630/2001007646-d.html

Griffiths J 2008 A brief history of mass spectrometry *Anal. Chem.* **80** 5678–83

Harrison A G 2018 *Chemical Ionization Mass Spectrometry* 2nd edn (Boca Raton, FL: CRC Press)

Houk R S, Fassel V A, Flesch G D, Svec H J, Gray A L and Taylor C E 1980 Inductively coupled argon plasma as an ion source for mass spectrometric determination of trace elements *Anal. Chem.* **52** 2283–9

Mason E A and McDaniel E W 1988 *Transport Properties of Ions in Gases* (New York: Wiley)

Nolting D, Malek R and Makarov A 2019 Ion traps in modern mass spectrometry *Mass Spectrom. Rev.* **38** 150–68

Tswett M 1906 Adsorptionsanalyse und chrematographische methode. Anwendung auf die Chemie des Chlorophylls *Ber. der Dtsch. Bot. Ges.* **24** 384–93

Vikse K L and Scott McIndoe J 2018 Ionization methods for the mass spectrometry of organometallic compounds *J. Mass Spectrom.* **53** 1026–34

Zimmermann R and Hanley (Hg.) L 2021 *Photoionization and Photo-Induced Processes in Mass Spectrometry: Fundamentals and Applications* (Weinheim: Wiley-VCH)

www.ingramcontent.com/pod-product-compliance
Ingram Content Group UK Ltd.
Pitfield, Milton Keynes, MK11 3LW, UK
UKHW051848100926
8556IPUK00005B/521

9 780750 331579